Sustained Simulation Performance 2014

Michael M. Resch • Wolfgang Bez • Erich Focht •
Hiroaki Kobayashi • Nisarg Patel
Editors

Sustained Simulation Performance 2014

Proceedings of the joint Workshop on Sustained Simulation Performance, University of Stuttgart (HLRS) and Tohoku University, 2014

Springer

Editors
Michael M. Resch
Nisarg Patel
High Performance Computing Center
Stuttgart (HLRS)
University of Stuttgart
Stuttgart
Germany

Wolfgang Bez
NEC High Performance Computing
Europe GmbH
Düsseldorf
Germany

Erich Focht
NEC High Performance Computing
Europe GmbH
Stuttgart
Germany

Hiroaki Kobayashi
Cyberscience Center
Tohoku University
Sendai
Japan

Front cover figure: Schematic view of Integrated Earthquake Simulation of Tokyo Metropolis Earthquake for seismic response analysis. The number of buildings analyzed exceeds 10,00,000. Illustrated by Muneo Hori, Earthquake Research Institute, The University of Tokyo, Tokyo, Japan

---

ISBN 978-3-319-10625-0 ISBN 978-3-319-10626-7 (eBook)
DOI 10.1007/978-3-319-10626-7
Springer Cham Heidelberg New York Dordrecht London

Library of Congress Control Number: 2014956566

Mathematics Subject Classification (2010): 68Wxx, 68W10, 68Mxx, 68U20, 76-XX, 86A10, 70FXX, 92Cxx, 65-XX

Printed on acid-free paper

Springer is part of Springer Science+Business Media (www.springer.com)

Preface

The field of high-performance computing is currently witnessing a significant shift of paradigm. Ever larger raw number crunching capabilities of modern processors are in principle available to computational scientists. Imperative knowledge of efficiently exploiting modern processors and performance achievements in the scientific community is growing by leaps and bounds.

On the other hand, many areas of computational science have reached a saturation in terms of problem size. Scientists often do no longer wish to solve larger problems. Instead they wish to solve smaller problems in a shorter time. The current architectures, however, are much more efficient for larger problems than they are for the more relevant smaller problems.

This series of workshops focuses on *sustained simulation performance*, i.e. high-performance computing for real application use-cases, rather than on peak performance, which is the scope of artificial problem sizes. The series was established in 2004, initially named Teraflop Workshop, and renamed to Workshop for Sustained Simulation Performance in 2012. In general terms, the scope of the workshop series has shifted from optimization for vector computers as the NEC SX-8, through efficient usage of large-scale systems, including NEC SX-9 but also cluster installations, to emphasis on future challenges, productivity and feasibility of current and future high-performance computing systems.

This book presents the combined results of the 18th and 19th installments of the series. The 18th workshop was held at the High-Performance Computing Center, Stuttgart, Germany, in October 2013. The 19th workshop was held in March 2014 at Sendai, Miyagi, Japan, and organized jointly with the University of Tohoku, Sendai, Japan.

The topics studied by the contributed papers include application driven approach towards future of HPC systems (Part I), framework analysis and scalability, exploitation of performance and productivity of the modern and existing hardware architectures (Part II), and application use-cases studies in interdisciplinary field (Part III).

We would like to thank all the contributors and organizers of this book and the Sustained Simulation Performance project. We especially thank Prof. Hiroaki

Kobayashi for the close collaboration over the past years and are looking forward to intensify our cooperation in the future.

Stuttgart, Germany Nisarg Patel
Stuttgart, Germany José Gracia
Stuttgart, Germany Michael Resch
August 2014

Contents

Part I Sustainability of Future HPC Systems: Application Driven Challenges

Feasibility Study of a Future HPC System for Memory-Intensive Applications: Final Report 3
Hiroaki Kobayashi
1 Introduction 3
2 System Architecture 7
3 Performance Estimation 10
4 Summary 15
References 16

The GASPI API: A Failure Tolerant PGAS API for Asynchronous Dataflow on Heterogeneous Architectures 17
Christian Simmendinger, Mirko Rahn, and Daniel Gruenewald
1 Introduction 17
2 GASPI Overview 19
2.1 History 19
2.2 Goals 20
3 The GASPI Concepts 21
3.1 GASPI Execution Model 21
3.2 GASPI Groups 21
3.3 GASPI Segments 22
4 GASPI One-Sided Communication 23
4.1 Basic Calls 24
4.2 Weak Synchronization 25
4.3 Extended Calls 26
5 GASPI Passive Communication 27
6 GASPI Global Atomics 29
7 GASPI Collective Communication 29

8 GASPI Failure Tolerance 31
8.1 GASPI Timeouts 31
8.2 GASPI Error Vector 31
Conclusion 31
References 32

Characteristic Analysis of Applications for Designing a Future HPC System 33
Osamu Watanabe, Takashi Soga, Youichi Shimomura, and Akihiro Musa
1 Introduction 34
2 Social and Scientific Challenges 35
2.1 Natural Disaster Mitigation 35
2.2 High Productivity Engineering 35
3 Application Requirements for the Future System 37
4 Performance Estimation on our Designed System 40
5 Potential of Overcoming the Challenges by Using our Designed System 43
5.1 Natural Disaster Mitigation 43
5.2 High Productivity Engineering 43
6 Summary 44
Appendix 45
References 45

Enhancing High Performance Computing with Cloud Concepts and Technologies 47
Bastian Koller and Michael Gienger
1 Introduction 47
2 Current Situation in HPC 48
3 High Performance Computing and/or Clouds 49
3.1 High Performance Computing Compared with Clouds 49
3.2 Complementary Use of HPC and Cloud 50
4 Cloud Based Access to HPC: Fortissimo as an Example 51
4.1 Introducing the Fortissimo Project 51
4.2 Realizing the One-Stop-Shop 51
5 The Road to Further HPC-Cloud Solutions 54
Conclusions 55
References 56

SX-ACE, Brand-New Vector Supercomputer for Higher Sustained Performance I 57
Shintaro Momose
1 Introduction 57
2 Architecture of SX-ACE 59
3 Implementation 62
4 Performance Evaluation 64

Conclusions 66
References 67

SX-ACE, the Brand-New Vector Supercomputer for Higher Sustained Performance II 69
Noritaka Hoshi and Shintaro Momose
1 Introduction 69
2 Concept of Design 70
2.1 Big Core Concept 70
2.2 Reduction of Power and Space 71
3 Architecture Overview 72
4 Implementation 74
5 Performance Evaluation 76
Conclusion 78
References 79

Feasibility Study of a Future HPC System for Memory Intensive Applications: Conceptual Design of Storage System 81
Ken'ichi Itakura, Akihiro Yamashita, Koji Satake, Hitoshi Uehara, Atsuya Uno, and Mitsuo Yokokawa
1 Introduction 81
2 Objectives 82
2.1 Design Cycle 82
2.2 Requirements from Applications 82
3 Design Concept and Result 83
4 Storage System Performance 85
5 Summary 87
References 88

Part II Exploitation of Existing HPC Systems: Potentiality, Performance and Productivity

Designing an HPC Refactoring Catalog Toward the Exa-scale Computing Era 91
Ryusuke Egawa, Kazuhiko Komatsu, and Hiroaki Kobayashi
1 Introductions 91
2 Performance Portability of HPC Applications 92
3 Designing an HPC Refactoring Catalog 93
3.1 Design Concepts 93
3.2 Current Status of the HPC Refactoring Catalog 96
3.3 Ongoing and Future Work 97
Conclusions 98
References 98

Endorsing Supercomputing Applications to Java Language ... 99
Alexey Cheptsov and Bastian Koller
1 Introduction ... 99
2 Related Work ... 101
2.1 MPI Bindings for Java ... 101
2.2 Native C Implementations of MPI ... 102
2.3 Non-MPI Based Approaches ... 102
3 Design and Implementation ... 104
3.1 Objectives ... 104
3.2 Architecture ... 105
3.3 Configuration and Running ... 107
4 Performance Evaluation ... 109
4.1 Basic Benchmarks ... 109
4.2 Pilot Application Scenario: Random Indexing Over Large Text Sets ... 113
5 Future Work ... 114
Conclusion ... 116
References ... 116

Performance Evaluation of an OpenMP Parallelization by Using Automatic Parallelization Information ... 119
Kazuhiko Komatsu, Ryusuke Egawa, Hiroyuki Takizawa, and Hiroaki Kobayashi
1 Introduction ... 119
2 OpenMP Parallelization by Using Automatic Parallelization Information ... 121
3 Performance Evaluation ... 122
3.1 Experimental Environments ... 122
3.2 Performance of OpenMP Codes Parallelized by using Automatic Parallelization Information ... 123
Conclusions ... 125
References ... 126

EXTOLL and Data Movements in Heterogeneous Computing Environments ... 127
Holger Fröning
1 Introduction ... 127
2 EXTOLL ... 128
2.1 Communication Engines ... 130
2.2 Key Performance Characteristics ... 132
2.3 Additional Reading ... 132
3 Global GPU Address Spaces ... 133
3.1 GPUs and Accelerated Clusters ... 133
3.2 A Thread-Collaborative Communication Model ... 134
3.3 Key Performance Characteristics ... 136
3.4 Additional Reading ... 136

4 Related Work ... 136
Conclusion ... 137
References ... 138

Requirements for Modern Network Infrastructures ... 141
Jens Aßmann, Alexander Kiontke, and Sabine Roller
1 Motivation ... 141
1.1 MPLS Traffic Engineering in OSPF Networks a Combined Approach ... 142
1.2 Enabling Software Defined Network (SDN) in Old School Networks with Software-Controlled Routing Protocols ... 142
2 Requirements for Modern Network Development at the University ... 143
2.1 Collision Domain ... 143
2.2 Routing in the Core ... 144
2.3 Routing with Redundant ISP Connection ... 146
2.4 Optical Fibre ... 147
2.5 Optical Fibre with MPLS ... 147
Conclusion ... 149
3 Further Work ... 149
References ... 149

Interconnection Network: Design Space Exploration of Network for Supercomputers ... 151
Kentaro Sano
1 Introduction ... 151
2 Assumption for Design Space Exploration ... 152
3 Preliminary Comparison Among Possible Topologies ... 153
4 Detailed Evaluation ... 158
Conclusions ... 160
References ... 161

Part III Computational Approach Towards Engineering and Multi-Physics Applications

Experiences in Developing HPC Software with Portable Efficiency ... 165
Daniel Friedrich Harlacher, Harald Klimach, and Sabine Roller
1 Introduction ... 165
2 Building Blocks in HPC Software Design ... 167
2.1 Implementation Language ... 167
2.2 Portability ... 168
2.3 Ease of Use ... 169
2.4 Maintaining a Scientific HPC Application ... 170
Conclusions ... 171
References ... 171

Petascale Computations for Large-Scale Atomic and Molecular Collisions 173
Brendan M. McLaughlin and Connor P. Ballance
1 Introduction 173
2 Parallel R-matrix Photoionization 175
3 Scalability 175
4 X-ray and Inner-Shell Processes 178
5 Heavy Atomic Systems 180
5.1 Kr and Xe Ions 180
5.2 Tungsten (W) Ions 182
6 Future Directions and Emergence of GPUS 184
References 185

FPGA-Based Scalable Custom Computing Accelerator for Computational Fluid Dynamics Based on Lattice Boltzmann Method 187
Kentaro Sano
1 Introduction 187
2 Tightly-Coupled FPGA Cluster for Scalable Custom Computing 189
2.1 Architecture of Tightly-Coupled FPGA Cluster 189
2.2 Design and Implementation of a Cluster Node 190
2.3 Acceleration Framework on FPGA 192
3 Case Study: Custom Computing with Lattice Boltzmann Method 193
3.1 Lattice Boltzmann Method 193
3.2 Architecture for Stream Computation 194
3.3 PE Design for Fully-Streamed Computation 195
4 Performance Evaluation 196
4.1 Implementation of PEs 196
4.2 Resource Consumption 197
4.3 Computational Performance 198
Conclusions 200
References 200

Application of HPC to Earthquake Hazard and Disaster Estimation 203
Muneo Hori, Tsuyoshi Ichimura, Maddegedara L.L. Wijerathne, and Kouhei Fujita
1 Introduction 203
2 Overview of HPC Application 204
2.1 Capability Computing 204
2.2 Capacity Computing 205
3 Structure Seismic Response Analysis 206
3.1 Fault-Structure System of Nuclear Power Plant 206
3.2 Reinforced Concrete Pier 209
4 Urban Area Seismic Response Analysis 212
4.1 Overview of Urban Area Seismic Response Analysis 212
4.2 Partial Reproduction of 2011 Great East Japan Earthquake Disaster 214

4.3 Partial Estimation of Tokyo Metropolis Earthquake 216
Conclusion 219
References 219

Geometry Dependent Computational Study of Patient Specific Abdominal Aortic Aneurysm 221
Nisarg Patel and Uwe Küster
1 Introduction 221
2 Image Modeling 223
2.1 Image Acquisition and Segmentation 223
2.2 Image Processing 223
3 Computational Modeling 228
3.1 Finite Element Model 230
3.2 Fluid Simulation Model 230
4 Results 233
References 237

Part I
Sustainability of Future HPC Systems: Application Driven Challenges

Feasibility Study of a Future HPC System for Memory-Intensive Applications: Final Report

Hiroaki Kobayashi

Abstract In the last 2 years, we have been involved in a project entitled, "a feasible study of a future HPC system for memory-intensive applications." In this project, we have analyzed some representative applications that need exascale computing around 2020, and clarified design specifications to develop a high-end computing system that will become available around 2018 and be best suited for these applications. This article reports results of a conceptual design and performance estimation of the system obtained through the project.

1 Introduction

According to the projection of the trend in the top 500 ranking to the future, many people expect that an exa-flop/s system will be available around 2019, and ranked at No.1 in that time frame [5]. Under such expectations of the HPC community around the world, US, Europe, China and Japan started several HPC strategic programs for targeting at realization of exascale systems around 2020.

In Japan, MEXT (Ministry of Education, Culture, Sports, Science and Technology) organized a committee to discuss the HPC policy of Japan for the next 5 to 10-year research and development on national leading-supercomputers in the era after K-computer, which was the first 10-peta flop/s LINPACK system in 2011. During the discussion, the committee decided to start a program entitled *Feasibility Study of Future HPCI systems* last year. The objectives of this program is to

- discuss future high-end systems capable of satisfying the social and scientific demands for HPC in the next 5–10 years in Japan, and
- investigate hardware and software technologies for developing future high-end systems available around year 2018 that satisfy the social and scientific demands.

In this project, three teams, which are University of Tokyo with Fujitsu (Project Leader: Professor Yutaka Ishikawa), University of Tsukuba with Hitachi (Project

H. Kobayashi (✉)
Tohoku University, Sendai 980-8578, Japan
e-mail: koba@isc.tohoku.ac.jp

M.M. Resch et al. (eds.), *Sustained Simulation Performance 2014*,
DOI 10.1007/978-3-319-10626-7_1

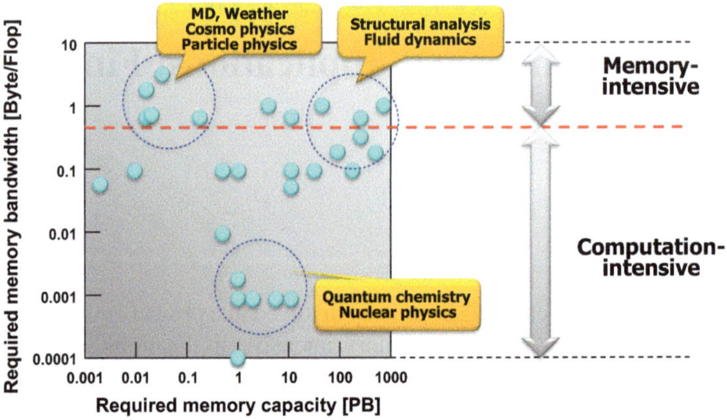

Fig. 1 Memory requirements of applications

Leader: Professor Mitsuhisa Sato), and Tohoku University with NEC (Project Leader: Hiroaki Kobayashi), started the feasibility studies of high-end computing systems in the exa-scale computing era as a 2-year national project in 2012. In this article, we will present our system design approach to exascale computing, especially for memory-intensive applications.

The system design philosophy of the Tohoku University team is to increase the productivity of high-performance computing. In the last decade, the peak performance of high-end computing systems is incredibly boosted by aggregating a huge number of nodes, each of which consists of multiple fine-grain cores, because the LINPACK benchmark for the top 500 ranking is computation-intensive, and inflating peak computing performance is a key factor to get a higher position in ranking rather than enhancing the memory throughput. However, according to the application development roadmap report summarized in Japan in 2012 [3], many important applications in the wide variety of science and engineering fields are memory-intensive and need 0.5 B/F or more for HPC system, as shown in Fig. 1. Here, B/F stands for *bytes per flop* that can be defined as a ratio of a memory throughput in bytes/s to a computing performance in flop/s of a HPC system. Therefore, if we continue to develop high-end computing systems by concentrating on increasing flop/s rates, simply targeting toward exa-flop/s in 2020, rather than memory bandwidth, their applicable areas are becoming narrowed, i.e., there will be a high probability that only few percentage of peak performance of exa-flop/s would be effective in the execution of practical applications, because lots of arithmetic units are stalled due to waiting for the arrival of data and end up wasted during their executions. Therefore, we rethink the design of high-performance computing systems with the quality of parallel processing, not the quantity of parallel processing for the era of exascale computing around 2020. Our key message is to *realize 100× more sustained performance with 10× peak performance for*

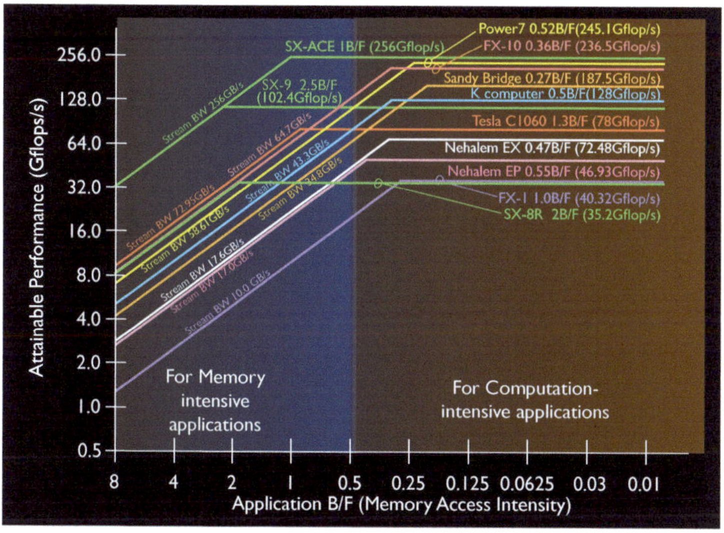

Fig. 2 Attainable performances of HPC processors in the roofline model

memory-intensive applications, compared with K-computer!, by exploiting sleeping computing capability with increasing memory, network, and I/O throughputs

To make HPC much more productive, we have to improve the efficiency of computing as much as possible. The efficiency is defined as the ratio of sustained performance of a real application to the peak performance of a HPC system. To this end, we have two design policies for future HPC systems. One is to make memory-throughput balanced with floating point performance to keep a high bytes per flop rate, i.e. 10× or more compared with those of competitive HPC systems. The other is to make processing cores much more coarse-grain high performance to avoid excessive parallel processing by using fine-grain low performance cores of the current trend in processor design. Figure 2 shows attainable performances of several representative HPC processors in the roofline model [4] as a function of application B/F, which is defined by the number of bytes for memory accesses to the number of floating point operations of a highest kernel of an application. The roofline model suggests that peak performances of individual processors make sense only for applications with an application B/F of 0.25 or lower, and for applications with 0.25 or larger, i.e., memory-intensive applications, their performances are severely limited by memory throughput. Therefore, as the throughput of the commodity-based memory subsystem improves very slowly compared with the performance in flop/s of a HPC system, B/F rates of future HPC systems are getting small, and their applicable areas are getting narrowed accordingly. As a result, we have to keep a memory throughput of a HPC system balanced with a flop/s performance to make the HPC system applicable to and productive for a wide variety of science and engineering applications.

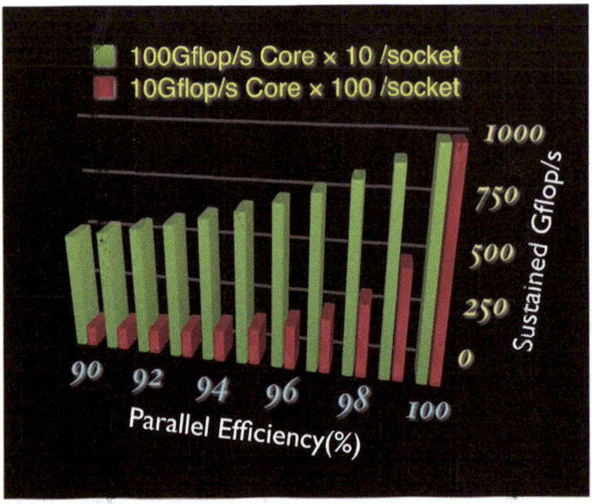

Fig. 3 Sustained performance of large-grain cores and fine-grain cores

The second point for the design of future productive HPC systems is to make parallel processing granularity larger to reduce the size of parallel processing. Figure 3 shows a comparison in sustained performance between two-type implementations of a 1 Tflop/s processor, one is ten 100 Gflop/s large-grain cores and the other is hundred 10 Gflop/s fine-grain cores. As the number of cores increases, overheads due to communications and synchronizations become significant. Therefore, a system with a smaller number of large-grain cores leads to a higher sustained performance even with a lower parallel efficiency as Amdahl's law suggests. Of course, larger cores have to achieve a higher efficiency with intra-core level vectorization and parallelization supported by a high memory throughput.

Based on the design polices discussed the above, we have come up with goals and their approaches in the design of a future HPC system that realizes highly productive HPC, especially for memory-intensive applications, as shown in Fig. 4. In order to realize a high memory bandwidth balanced with a peak flop/s performance, we design a memory subsystem with a B/F rate of 1 up to 2 by aggressively introducing innovative device technologies such as the 2.5D interposer technology, the 3D die-stacking technology and their combination. A B/F rate of up to 2 is very challenging design specification and is 10× or more than the expectation that is currently discussed on design of future HPC systems. In addition design of the memory subsystem, designs of highly efficient large-grain processing cores, sockets, nodes, and network are also important to make parallel processing much more efficient and productive. To improve vector processing capability, we design an advanced vector architecture with a large on-chip vector load/store buffer to reduce penalties for handling short-vectors and indirect memory accesses, and with these large-grain vector processing cores, we construct a large node with a high-bandwidth, large

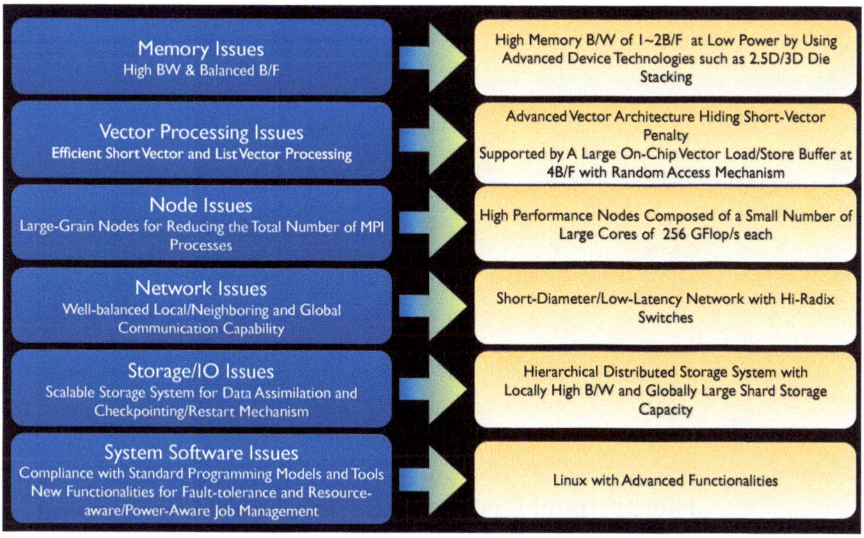

Fig. 4 Goals and approaches of a future HPC system for memory-intensive applications

shared memory. We try to reduce the total number of nodes connected via a network to satisfy a certain level of sustained performances that target applications need.

The design of a storage system is also important because there is an increasing demand to handle big data in HPC such as data assimilation for the atmosphere-ocean coupled simulation. In addition, the storage system plays an important role to support a checkpoint/restart mechanism that is mandatory to make the system much more dependable. Therefore, our approach to the scalability storage system for a large data assimilation and efficient checkpointing/restarting is to design a hierarchical distributed storage system with locally high bandwidth and globally large shared storage capacity. The last but not least issue is system software design. We investigate a Linux based operating system with several advanced functions that provide a standard programming environment specially designed for our vector-parallel processing mechanism under the high fault-tolerance and effective resource-aware/power-aware job managements.

2 System Architecture

Figure 5 shows the overview of a conceptual design of our system. A CPU socket consists of four vector-processing cores and a 32 MB VLSB (vector load store buffer). Each core has a performance of 256 Gflop/s and is connected to the VLSB at a bandwidth of up to 1 TB/s, resulting in a B/F rate of 4 per core and a peak

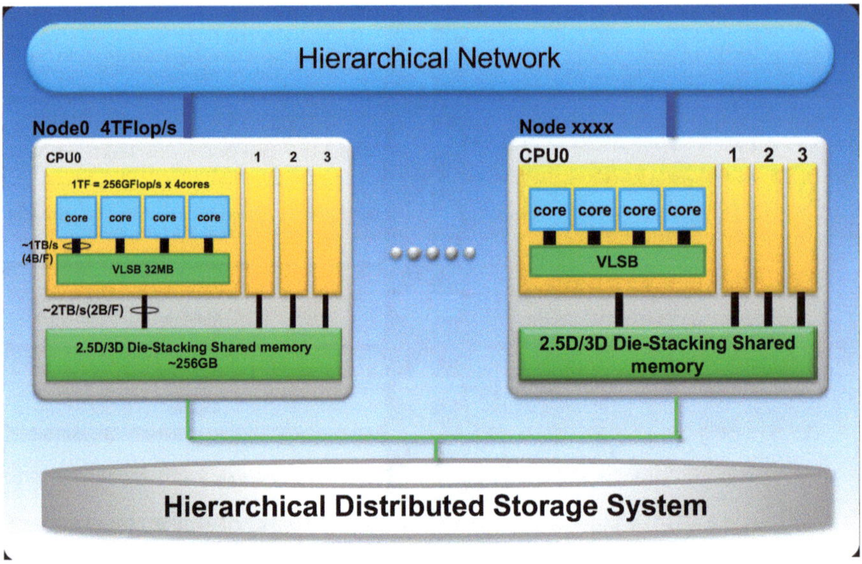

Fig. 5 Overview of the system architecture

CPU performance of 1 Tflop/s. VLSB is a software-controllable on-chip buffer to support vector load/store operations and works very effective when they have the locality of reference in their vector processing. *Software-controllable* means that VLSB can selectively hold vector data with the high locality only to reduce the pollution of VLSB by vector data with the low locality. VLSB can also effectively assist gather/scatter operations to reduce the latency of list vector processing.

A node consists of four CPU sockets and a shared memory. The shared memory has four channels to connect CPU sockets at a bandwidth of up to 2 TB/s. Therefore, the node B/F is also up to 2. To realize such a high memory bandwidth, we design the memory subsystem with 2.5D and 3D die-stacking technologies. We design two types of the memory subsystem under the consideration of the tradeoff between performance and cost. One is the custom design and the other is the commodity-based design as shown in Fig. 6. The custom design is aggressive one to realize 2 B/F of the node. Figure 7 shows the design of the memory modules by using four stacked DRAM devices and custom designed memory controller on a Si interposer connected through TSVs (Through Silicon Vias). The commodity-based memory subsystem design is also examined to reduce the cost, while sacrificing performance. In the commodity-based design, we use Hybrid Memory Cube (HMC) devices of Micron[2] instead of DDR4 devices[1]. Both of them are expected to become available around 2018, however, HMC is more promising to obtain a high bandwidth of the memory subsystem at the level of 1 B/F, even though the cost of HMC is higher than that of DDR4 due to the emerging technology of 3D die stacking for high-bandwidth memory device design of HMC. The B/F of the commodity-based

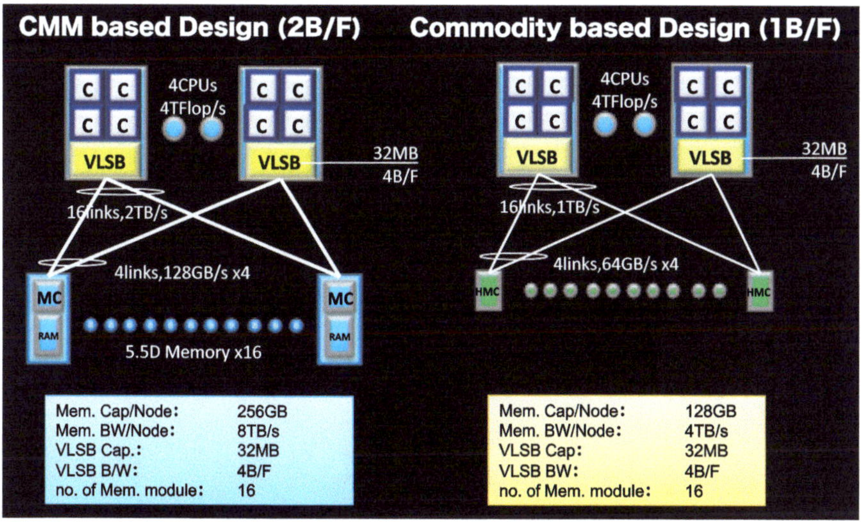

Fig. 6 A memory subsystem

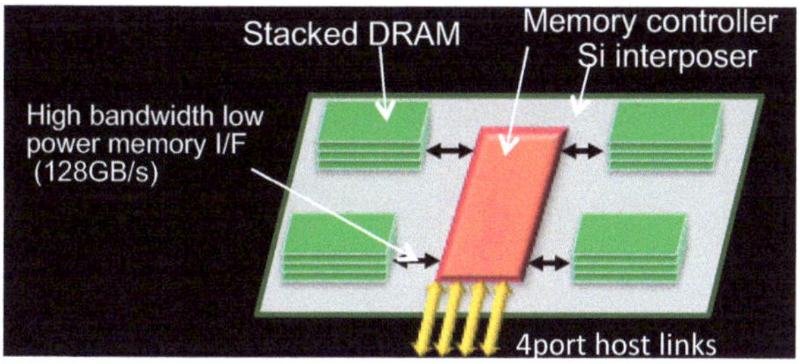

Fig. 7 Memory module design

design is half of that of the custom design, but we think that it is still 10x higher than that for the scalar-based architecture expected to become available around 2018.

We examined two topologies for the network system, which are fat tree and fat-tree torus hybrid (FTT-Hybrid) as shown in Fig. 8. As the preliminary evaluation in Fig. 9 suggests, Fat-tree has a shorter diameter but a relatively larger cable delay compared with FTT-Hybrid. FTT-Hybrid has a lower cable delay and lower cost, however, we have decided to use Fat-Tree because it also provides the easiness and flexibility of job scheduling when partitioning the system for scheduling a lot of small and/or medium sized jobs in addition to the advantages regarding its topology.

Figure 10 shows a hierarchical storage system. The storage system at the first layer is an SSD-based local storage to realize a high bandwidth for handling big data and checkpointing/restarting, and the second layer is composed of a lot of commodity hard disk drives as a large global storage to hold big data such as

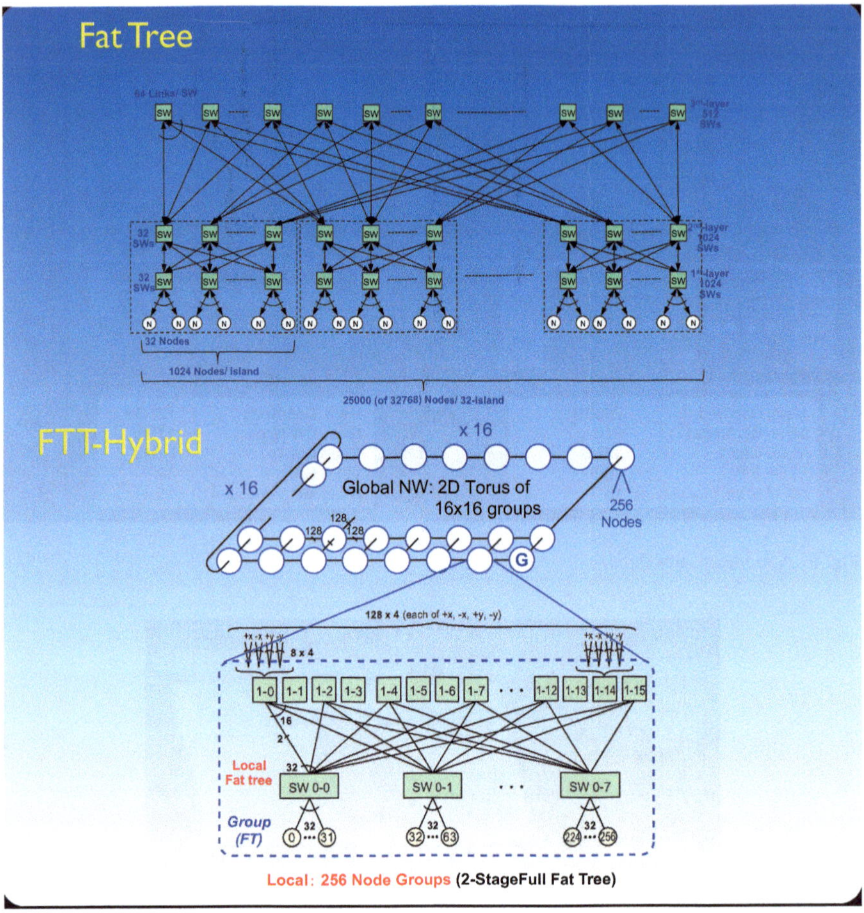

Fig. 8 Examined networks

data assimilation and check-pointed snapshots of memory images. The detailed discussion regarding the storage system will be available in [7] in this book.

3 Performance Estimation

To evaluate our design presented in Sect. 2, we have developed and analyzed target applications that will be needed around 2020 to satisfy the social and scientific demands for HPC at that time. In particular, after the 3.11 Great East Japan Earthquake in 2011, the people of Japan seriously concern about the secure and safe homeland against natural disasters. Therefore, we examine total ten applications that cover natural disaster prevention and mitigation areas such as earthquake, tsunami,

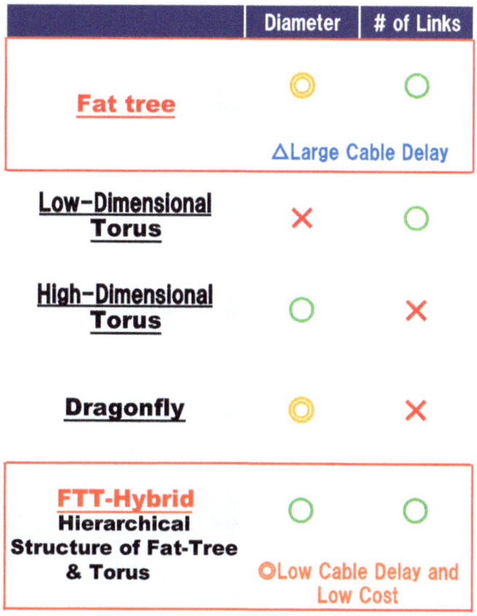

Fig. 9 Preliminary evaluation of network topologies

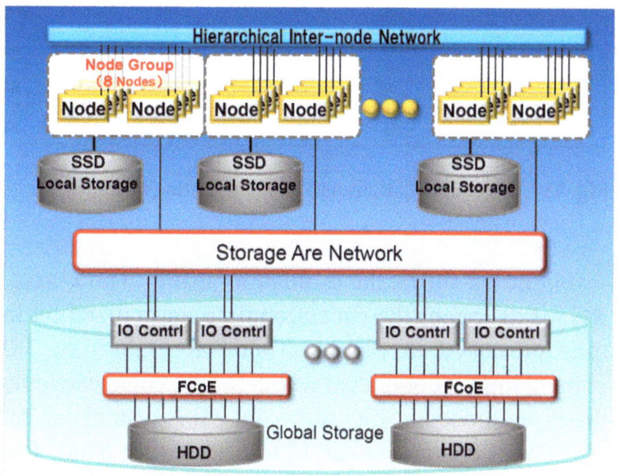

Fig. 10 A hierarchical storage system

typhoon and their compound phenomenon. In addition to applications for natural disasters prevention and mitigation, we also investigate some important applications in the engineering field. The details of our target applications analyzed in the project are available in [6] in this book.

Figure 11 summarize computation, memory, and I/O demands of our 14 target applications. All the applications are memory-intensive, which need B/F of 2 or more, up to over 9. They also need the exa-level computation that should be executed within several hours. By using these applications, we evaluate our designed system through a simulator developed. In the evaluation, we analyze two types of systems: the custom memory system with 2 B/F and the commodity memory system with 1 B/F. These two system have the same computing performance of a 100 Pflop/s equipped with a 250 PB storage and a 40 GB/s Fat-Tree network.

Application	B/F	Required Floating-Point Operations($\times 10^{18}$)	Total Memory Size	Total File I/O Size	Expected Execution Time
RSGDX	8.00	1.7	33 TB	26 TB	<24 Hours
Seism3D	2.14	870	2.8 PB	433 TB	<8 Hours
QSFDM_GLOBE	1.39	86.5	1.7 TB	284 GB	<12 Hours
MMA	5.56	1376	703 TB	16.2 PB	<10 Days
STOC-CADMAS	9.31	116	42.3 TB	4.1 PB	<24 Hours
TUNAMI-CADMAS	9.31	19	0.6 TB	378 TB	<15 Hours
MSSG	4.00	720	175 TB	1.6 PB	<6 Hours
CDA	4.24~4.99	2	18 TB	9 TB	<4 Hours
GSM	4.31	0.4	21 TB	4 TB	<1.4 Hours
Compound Disaster Analysis (1000 Cases)	2.14~9.31	39×1000	85.7 TB	832 TB ×1000	<4 months (=3-hour × 1000-cases)
Numerical Turbine (bunch of 20-case jobs)	2.33	259	163.5 TB	166 TB	<20 Hours
BCM	5.47	33	390 TB	338 GB	<0.5 Hours
BCM-LEE (bunch of 10-case jobs)	4.55	35000	688 TB	72 TB	<60 Hours
LGAsxmk54v	1.64	3300	880 TB	23 MB	<24 Hours

Fig. 11 Target applications and their computing demands

Figure 12 shows the evaluation results. The x-axis is for application names and the y-axis means the execution time normalized by individual expected execution time presented in Fig. 11. Therefore, values of 1.0 or smaller mean that the computation is completed within the expected execution time of that application. As the experimental results show, the 2 B/F system can satisfy the computing demands of 13 applications out of 14. On the other hand, the 1 B/F system competed six applications within the expected execution time, but even for the applications that the 1 B/F system cannot complete within the expected time, the system only needs a 30 % longer execution time on average than the expectation. We think that the 2 B/F system is the best solution to highly-productive HPC for memory-intensive applications in the time-frame of 2018–2020, however, as our cost and power consumption estimation suggests that the 1 B/F system can achieve a 20 % reduction in power consumption and a 30 % in manufacturing cost, the 1 B/F also shows a good trade-off between cost and performance.

The readers of this article may think that a 100 Pflop/s performance of our base design is quite lower than the expectation for the exa-scale and/or exa-flop/s computing. To answer this question, we examine a *most-likely commodity-system* available around 2018 in comparison with our designed system. Figure 13 shows our assumption of the commodity-based system to be available in 2018. The

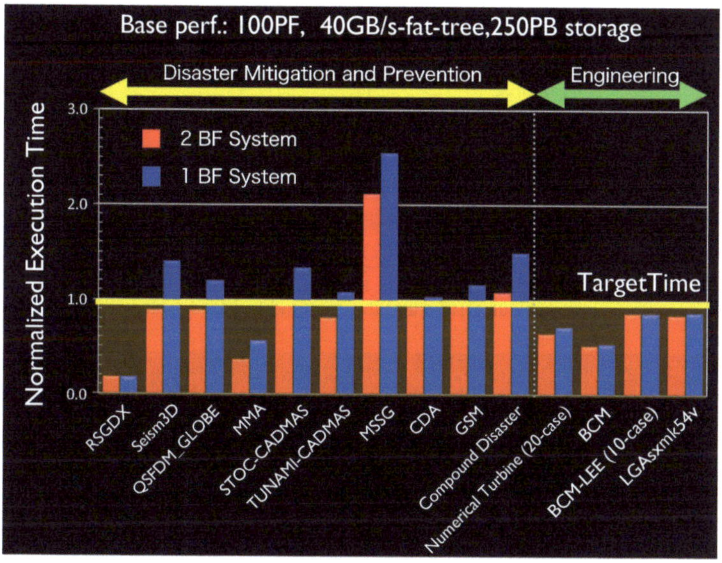

Fig. 12 Evaluation results

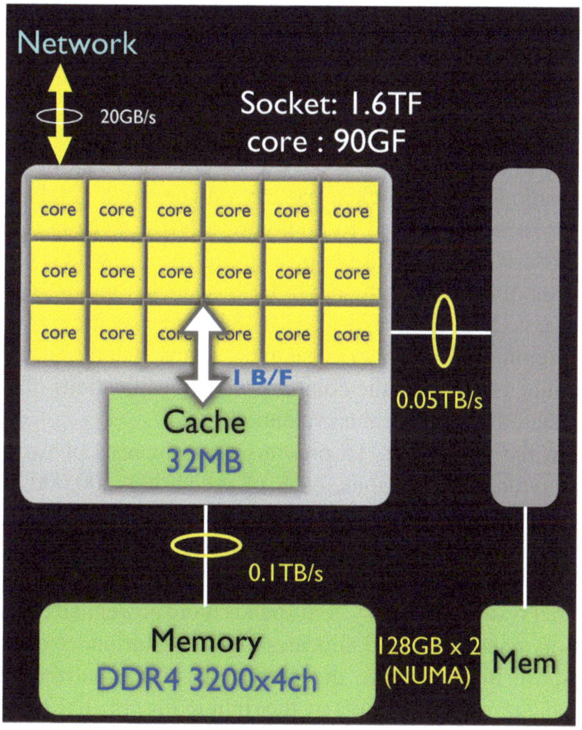

Fig. 13 Commodity-based system configuration

	Our System	Commodity-Based	Ratio
Tread/proc	4	4	
PeakPF/proc	1TF	0.36TF	2.8x
PeakPF/100,000proc	100PF	35.84PF	2.8x
Nodes /100,000proc	25,000	11,111	2.3x
Total MemBW	200PB/s	3.4PB/s	59x
Total Cache Capacity	3.2TB	0.8TB	4x

Fig. 14 Comparison between our system and the commodity-based system

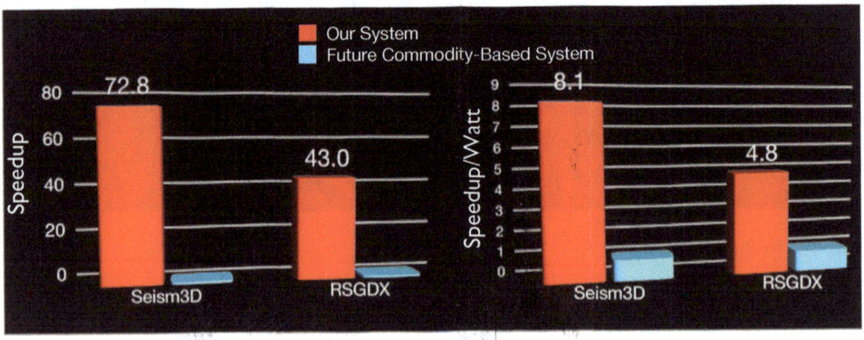

Fig. 15 Performance evaluation of Seism3D and RSGDX

specification and organization are derived based on the trend in the recent Xeon processor generations. Each socket has 18 cores and a total performance of 1.6 Tflop/s. Two sockets are used to configure a NUMA (Non-Uniform Memory Architecture) node with a 3.3 Tflop/s peak performance, which are connected to DDR4 memory modules at a bandwidth of 0.2 TB/s, resulting in 0.1 B/F. The power consumption of a node is estimated to be 0.4KW because the power consumption of a node in a commodity cluster system is almost fixed across the processor generations. Figure 14 presents the comparison in specifications of our system and the commodity-based system for the evaluation.

The graph on the left in Fig. 15 presents performances of two systems when executing three typical applications, each of which has 100,000 MPI processes. As the graph shows, our system outperforms the commodity-based system with 43× to 73× speedups. Therefore, our system successfully achieves a necessary sustained performance with the lower number of MPI processes compared with the commodity-based system. As our system spends more power and cost to increase the sustained memory bandwidth, and allocates higher performance cores to individual MPI processes, we also compare the performances normalized by their power consumptions, and present the results on the right in Fig. 15. The figure clarifies that our system achieves a significant improvement in speedup per watt, 4.8× to 8.1×, which are more than the peak performance ratio of 2.8× between the two systems.

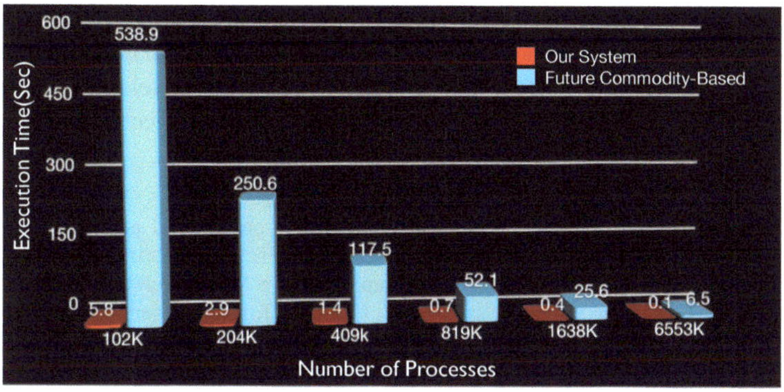

Fig. 16 Scalability analysis results (Seism3D)

Therefore, we can confirm that our system is also power-efficient in execution of memory intensive applications.

Many people said the massive parallel processing would be a solution, but for memory-intensive applications, this sometimes does not make sense. Figure 16 shows the results of the scalability analysis when increasing the number of MPI processes on both systems. The commodity-based system scales well until one million processes because the on-chip cache works. However, the scalability declines gradually afterward due to the limitation of the off-chip memory bandwidth, and the commodity system eventually needs 6.4 million MPI processes (cores) to achieve the same sustained performance of the our system. As the commodity-based system with 6.4 million cores has a peak performance of 3.2 Exa-flop/s, our 100 Peta-flop/s system with 2 B/F has a potential equivalent to the exa-flop/s system, and we believe that our system would be more productive because it can achieve the necessary sustained performance with the moderate number of high efficiency processing cores.

4 Summary

This article has presented the overview of our 2-year project entitled "Feasibility study of a Future HPC system for Memory-Intensive Applications." In this project, we much focused on increasing the efficiency of the HPC system for productive simulation. We have reviewed important applications to be solved around 2020 on so-called *exa-scale computing platforms,* and clarified the necessary hardware and system design technologies to realize a system that satisfies the requirements given by these applications.

Experimental results indicate that with the enhancement of the memory subsystem balanced with the improvement in floating point calculation performance to 2 B/F, we can complete exascale applications within their individual expected execution times even with 100 peta-flop/s peak performance, which would be equivalent to exa-flop/s peak performance of a commodity-based future system. There are many research issues to be solved before system developments, however, we continue to work together with many colleagues of this project to make our ideas into the reality.

Acknowledgements Many colleagues get involved in this project, and great thanks go to Dr. Y.Kaneda and Dr. K.Watanabe of JAMSTEC (Japan Agency for Marine-Earth Science and Technology) as co-leaders of the application group, Professor M. Yokokawa of Kobe University, Associate Professors H. Takizawa, K. Sano and R. Egawa of Tohoku University, and Dr. K.Itakura of JAMSTEC as sub-leaders of the architecture group, Professor M.Koyanagi of Tohoku University as the leader of the 2.5D/3D device group, and Ms. Y.Hashimoto of NEC as the leader of the NEC application, system and device design groups. This project is supported by Ministry of Education, Culture, Sports, Science and Technology of Japan.

References

1. Global Standard for the Microelectronics Industry. http://www.jedec.org/category/technology-focus-area/main-memory-ddr3-ddr4-sdram (2014)
2. Hybrid Memory Cube Consortium. http://www.hybridmemorycube.org (2014)
3. MEXT HPC Task Force: In: Japanese Report on Application (R&D). Roadmap for Exascale Computing (2012)
4. Samuel, W.: Roofline: an insightful visual performance model for floating-point programs and multicore architectures. Lawrence Berkeley National Laboratory. http://escholarship.org/uc/item/5tz795vq (2009)
5. Top 500 Supercomputer Sites. http://www.top500.org/ (2014)
6. Watanabe, O., et al.: Characteristic analysis of applications for designing a future HPC system. In: Sustained Simulation Performance 2014. Springer, Berlin (2014)
7. Yokokawa, M., et al.: Feasibility study of a future HPC system for memory intensive applications - conceptual design of storage system. In: Sustained Simulation Performance 2014. Springer, Berlin (2014)

The GASPI API: A Failure Tolerant PGAS API for Asynchronous Dataflow on Heterogeneous Architectures

Christian Simmendinger, Mirko Rahn, and Daniel Gruenewald

Abstract The Global Address Space Programming Interface (GASPI) is a Partitioned Global Address Space (PGAS) API specification. The GASPI API specification is focused on three key objectives: scalability, flexibility and fault tolerance. It offers a small, yet powerful API composed of synchronization primitives, synchronous and asynchronous collectives, fine-grained control over one-sided read and write communication primitives, global atomics, passive receives, communication groups and communication queues. GASPI has been designed for one-sided RDMA-driven communication in a PGAS environment. As such, GASPI aims to initiate a paradigm shift from bulk-synchronous two-sided communication patterns towards an asynchronous communication and execution model. In order to achieve its much improved scaling behaviour GASPI leverages request based asynchronous dataflow with remote completion. In GASPI request based remote completion indicates that the operation has completed at the target window. The target hence can (on a per request basis) establish whether a one sided operation is complete at the target. A correspondingly implemented fine-grain asynchronous dataflow model can achieve a largely improved scaling behaviour relative to MPI.

1 Introduction

As the supercomputing community prepares for the era of exascale computing, there is a great deal of uncertainty about viable programming models for this new era. HPC programmers will have to write application codes for systems which are hundreds of times larger than the top supercomputers of today. It is unclear whether the two-decades-old MPI programming model [1], by itself, will make that transition gracefully. Despite recent efforts in MPI 3.0, which have significantly improved

---

C. Simmendinger (✉)
T-Systems Solutions for Research, Stuttgart, Germany
e-mail: christian.simmendinger@t-systems-sfr.com

M. Rahn • D. Gruenewald
Fraunhofer ITWM, Kaiserslautern, Germany
e-mail: mirko.rahn@itwm.fraunhofer.de; daniel.gruenewald@itwm.fraunhofer.de

M.M. Resch et al. (eds.), *Sustained Simulation Performance 2014*,
DOI 10.1007/978-3-319-10626-7_2

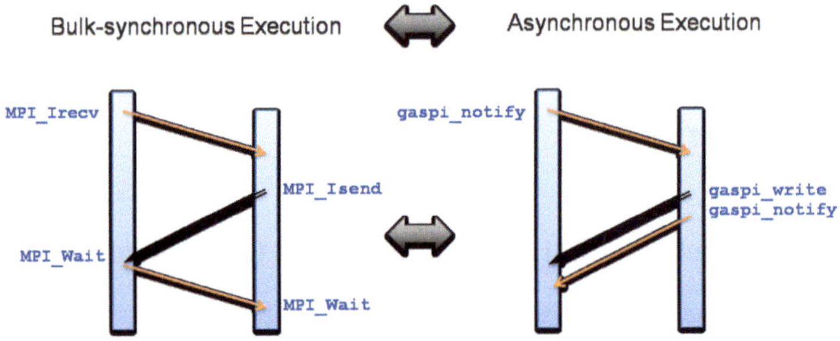

Fig. 1 MPI two-sided communication vs GASPI notify. An MPI_Irecv is complete at the receiver side whenever the corresponding wait returns. IN GASPI the one sided communication is complete at the receiver side, whenever the test for the corresponding notification successfully returns. Remark: The initial gaspi_notify here requests the subsequent write and notify, but this initial request is not mandatory

support for one-sided communication and an asynchronous execution model, MPI offers little support for asynchronous dataflow models.

We here present an alternative to the programming model of MPI. GASPI is a Partitioned Global Address Space (PGAS) API. In contrast to MPI, GASPI leverages one-sided RDMA driven communication with remote completion in a Partitioned Global Address Space. In GASPI data may be written asynchronously and one-sided, whenever it is produced, along with a corresponding notification. On the receiving side GASPI guarantees that data is locally available whenever this notification becomes locally visible. This mechanism enables fine-grain (request based) asynchronous dataflow implementations on top of the GASPI API (see also Fig. 1).

The notification mechanism in GASPI enables its users to reformulate existing bulk-synchronous MPI applications towards an asynchronous dataflow model with a very fine-grained overlap of communication and computation. GASPI leverages zero-copy asynchronous writes and reads with RDMA queues and aims at a minimum of communication overhead.

GASPI allows for a lot of flexibility in using its Partitioned Global Address Spaces (which in GASPI are called segments), but does not enforce a specific memory model. We remark that implementing a symmetric global memory management on top of the GASPI segments is straightforward. We have, however, opted against including this latter functionality in the core API, since enforcing a symmetric global memory management would make the GASPI API a bad match for e.g irregular problems.

GASPI allows its users to span multiple segments with configurable sizes and configurable participating ranks. GASPI also supports a variety of devices for its segments, like e.g GPGPU memory, main memory of Xeon Phi cards, main memory of host nodes or non-volatile memory. All these segments can directly read/write from/to each other—within the node and across all nodes.

With a growing number of nodes, failure tolerance becomes a major issue as machines expand in size. On systems with large numbers of processes, all non-local communication should be prepared for a potential failure of one of the communication partners. GASPI features timeouts for non-local functions, allows for shrinking or growing node sets and enables applications to recover from node-failures.

In contrast to other efforts in the PGAS community, GASPI is neither a new language (like e.g. Chapel from Cray [2], UPC [3] or Titanium [4]), nor an extension to a language (like e.g. Co-Array Fortran [5]). Instead—very much in the spirit of MPI—it complements existing languages like C/C++ or Fortran with a PGAS API which enables the application to leverage the concept of the Partitioned Global Address Space.

While other approaches to a PGAS API exist (e.g. OpenShmem or Global Arrays), they lack remote completion, support for multiple segments and failure tolerance.

2 GASPI Overview

2.1 History

GASPI inherits much of its design from the Global address space Programming Interface (GPI) [6, 7], which was developed in 2005 at the Competence Center for High Performance Computing (CC-HPC) at Fraunhofer ITWM. GPI is implemented as a low-latency communication library and is designed for scalable, real-time parallel applications running on cluster systems. It provides a PGAS API and includes communication primitives, environment run-time checks and synchronization primitives such as fast barriers or global atomic counters.

GPI communication is asynchronous, one-sided and, most importantly, does not interfere with the computation on the CPU. Minimal communication overhead can be realized by overlapping communication and computation. GPI also provides a simple, run-time system to handle large data sets, as well as dynamic and irregular applications that are I/O- and compute-intensive. As of today, there are production-quality implementations for ×86 and IBM Cell/B.E architectures.

GPI has been used to implement and optimize CC-HPC industry applications like the Generalized Radon Transform (GRT) method in seismic imaging or the seismic work flow and visualization suite PSPRO. Today, GPI is installed on Tier 0 supercomputer sites in Europe, including the HLRS in Stuttgart and the Juelich Supercomputing Centre.

The GPI library has yielded some promising results in a number of situations. In particular, GPI outperforms MPI in significant low-level benchmarks. For process to process communication, GPI asynchronous one-sided communication, as opposed to both MPI one-sided communication and MPI bulk-synchronous two sided-communication, delivers full hardware bandwidth. As a function of message size, GPI reaches its peak performance much earlier than MPI.

GPI has also shown excellent scalability in a broad spectrum of typical real world HPC applications like the Computational Fluid Dynamics (TAU code from the DLR) [8], or BQCD, a four dimensional nearest neighbor stencil algorithm. GPI has also been used in the implementation of fastest Unbalanced Tree Search (UTS) benchmark on the market [9].

In 2010 the request for a standardization of the GPI interface emerged, which ultimately lead to the inception of the GASPI project in 2011. The work was funded by the German Ministry of Education and Science and included project partners Fraunhofer ITWM and SCAI, T-Systems SfR, TU Dresden, DLR, KIT, FZJ, DWD and Scapos.

2.2 Goals

The GASPI project intends to drive the dissemination and visibility of the API by means of highly visible lighthouse projects in specific application domains, including CFD, turbo-machinery, weather and climate, oil and gas, molecular dynamics, as well as in the area of sparse and dense matrices. On the basis of these applications, requirements for the emerging standard have been analyzed. The following requirements have been derived:

- fine-grained control over one-sided asynchronous read/write operations
- collective operations (also on a subset of ranks)
- passive communication
- atomic counters

For the GASPI API we have set the following design goals:

- Extreme scalability, targeting both performance and minimal resource requirements
- Timeout mechanisms and failure tolerance
- Multi-segment support
- Dynamic allocation of segments
- Group support for collectives
- Large flexibility in runtime parameter configuration
- A slim, yet powerful core API providing a minimal still complete set of functionality
- A strong standard library extension, which takes care of convenience procedures.
- A maximum of freedom for the implementation

Ultimately the GASPI project aims at establishing a de-facto standard for an API for scalable, fault-tolerant and flexible communication in a Partitioned Global Address Space.

3 The GASPI Concepts

3.1 GASPI Execution Model

GASPI follows an SPMD (Single Program, Multiple Data) style in its approach to parallelism. Hence, a single program is started and initialized on all desired target computational units. How the GASPI application is started and initialized is not defined by the standard and is implementation specific.

GASPI provides the concept of ranks. Similarly to MPI each GASPI process receives a unique rank with which it can be identified during runtime. Ranks are a central aspect which allows applications to identify processes and to assign different tasks or data to the processing elements.

GASPI also provides the concept of segments. Segments are PGAS memory regions which can be globally available, to be written to or read from.

The GASPI API has been designed to coexist with MPI and hence in principle provides the possibility to complement MPI with a Partitioned Global Address Space. We note however, that while such an approach provides an opportunity for increased scalability, fault-tolerant execution will not be possible due to the corresponding limitations of MPI. GASPI aims at providing interoperability with MPI in order to allow for incremental porting of such applications.

The start-up of mixed MPI and GASPI code is achieved by invoking the GASPI initialization procedure in an existing MPI program. This way, MPI takes care of distributing and starting the binary and GASPI just takes care of setting up its internal infrastructure.

GASPI provides high flexibility in the configuration of the runtime parameters for the processes and allows for a dynamic process set during runtime. In case of a node failure, a GASPI process can be started on a new host, freshly allocated or selected from a set of pre-allocated spare hosts. By providing a modified list of machines (in which the failed node would be substituted by the new host) the new GASPI process would register with the other processes and receive the rank of the failed GASPI process.

Similarly, in case of starting additional GASPI processes, these additional GASPI processes have to register with the existing GASPI processes.

3.2 GASPI Groups

Groups are sub-sets of processes identified by a sub-set of the total set of ranks. The group members have common collective operations. A collective operation is then restricted to the ranks forming the group. Each GASPI process can participate in more than one group.

A group has to be defined and declared in each of the participating GASPI processes. Defining a group is a two step procedure. An empty group has to be created first. Then the participating GASPI processes, represented by their rank, have to be attached. The group definition is a local operation. In order to activate the group, the group has to be committed by each of the participating GASPI processes. This is a collective operation for the group. Only after successful group commitment the group can be used for collective operations.

The maximum number of groups allowed per GASPI process is restricted by the implementation. A desired value can be passed to the GASPI initialization procedure.

In case of failure, where one of the GASPI processes included within a given group fails, the group has to be reestablished. If there is a new process replacing the failed one, the group has to be defined and declared on the new GASPI process(es). Reestablishment of the group is then achieved by re-commitment of the group by the GASPI processes which were still "alive" (functioning) and by the commitment of the group by the new GASPI processes.

3.3 GASPI Segments

GASPI does not enforce a specific memory model, like, for example, the symmetric distributed memory management of OpenSHMEM [10]. Rather GASPI offers PGAS in the form of configurable RDMA pinned memory segments. Since an application can request several PGAS segments symmetric, asymmetric or stack based memory management models can readily coexist.

Modern hardware typically involves a hierarchy of memory with respect to the bandwidth and latencies of read and write accesses. Within that hierarchy are non-uniform memory access (*NUMA*) partitions, solid state devices (*SSD*s), graphical processing unit (*GPU*) memory or many integrated cores (*MIC*) memory.

The GASPI memory segments can thus be uses as an abstraction, which represents any kind of memory level, mapping the variety of hardware layers to the software layer. A segment is a contiguous block of virtual memory. In the spirit of the PGAS approach, these GASPI segments may be globally accessible from every thread of every GASPI process and represent the partitions of the global address space (Fig. 2).

By means of the GASPI memory segments it is also possible for multiple memory models or indeed multiple applications to share a single Partitioned Global Address Space. The segments can be accessed as global, common memory, whether local—by means of regular memory operations—or remote, by means of the communication routines of GASPI. Allocations inside of the pre-allocated segment memory are managed by the application.

In a Partitioned Global Address Space, every thread can asynchronously read and write the entire global memory of an application. On modern machines with RDMA engines, an asynchronous PGAS programming model appears as a natural extension and abstraction of available hardware functionality.

For systems with DMA engines (such as tile architectures), this also holds true for a node-local level.

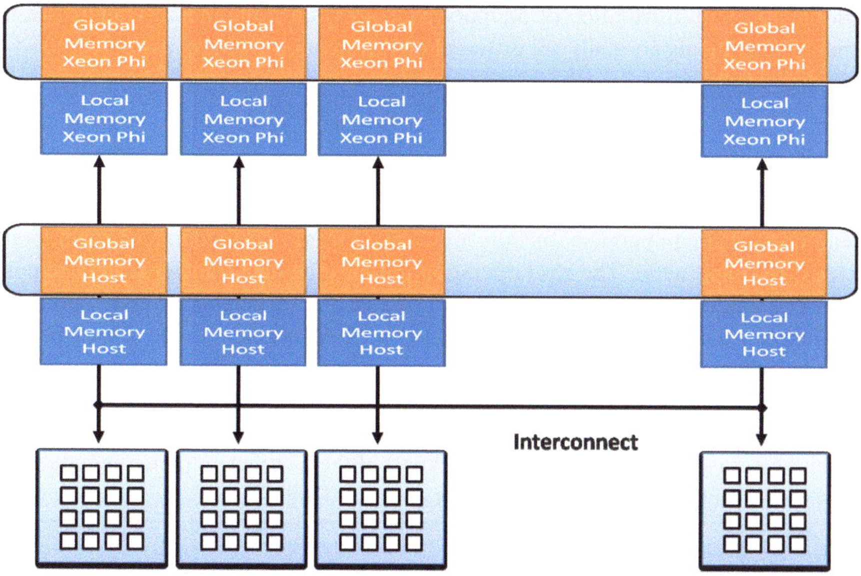

Fig. 2 Memory segments in a Partitioned Global Address Space. Here one segments is spawned across all Xeon Phi cards and a different segment is spawned across the ×86 host memory. The segments can directly read/write from/to each other—within the node and across all nodes

4 GASPI One-Sided Communication

One sided asynchronous communication is the basic communication mechanism provided by GASPI. Here one GASPI process specifies all communication parameters, both for the local and the remote side.

GASPI provides the concept of message queues. All operations placed on a certain queue q by one or several threads are finished after a single wait call for q has returned. These queues facilitate higher scalability and can be used as channels for different types of requests where similar types of requests are queued and then get synchronized together but independently from the other ones (separation of concerns, e.g. one queue for operations on data and another queue for operations on meta data). The several message queues guarantee fair communication, i.e. no queue should see its communication requests delayed indefinitely. Furthermore, the queue preserves the order of the messages on the local and the remote side if the remote rank of two messages is the same. The number of message queues and their size can be configured at initialization time, otherwise default values will be used. The default values are implementation dependent. Maximum values are also defined.

Memory addresses within the global partitioned address space are specified by the triple consisting of the rank, segment identifier and the offset. This prevents a global all to all distribution of memory addresses, since the memory addresses of memory segments normally are different on different GASPI processes,

4.1 Basic Calls

The one-sided communication calls are derived from two fundamental operations types: read and write. The read operation initiates the data transfer from a remote part of a PGAS segment to the local part of a PGAS segment. The write operation initiates the data transfer from a local part of a PGAS segment to the remote part of a PGAS segment.

Listing 1 shows an alltoall implementation with one-sided writes. This implementation performs an out-of-place matrix transpose of a distributed square matrix. Each process iProc hosts one row (src[iProc]) of the distributed matrix and also one row (dst[iProc]) of the transposed matrix.

```
let nProc be the number of processes;
let iProc be the unique local process id;
let src[iProc] be an array of size nProc;
let dst[iProc] be an array of size nProc;

foreach process p in [0,nProc):
  write src[iProc][p] into dst[p][iProc];
  //    ^^^^^^^^^^^^   ^^^^^^
  //  | local | remote if p != iProc

work()
// do some work along the data transfer

wait for the completion of the writes;
// now this process sent the data

barrier;
// now all processes sent the data
```

Listing 1 Alltoall with one-sided writes

Each process loops over its row entries src[iProc][p] and initiates the data transfer to the corresponding row entry src[p][iProc] of process p. The communication requests are posted to one of the communication queues and are asynchronously processed by the network infrastructure.

Parallel to the data transfer, a local task (work) can be executed in order to overlap the communication with the computation. Upon completion of the local task, the local process needs to know whether the source buffers can be reused again. Hence a wait call has been invoked on the respective queue. This is a blocking call. However, if the computation of the local task has been sufficiently long, the data transfer has finished and the wait call returns immediately.

4.2 Weak Synchronization

One-sided communication procedures have the characteristics that the entire communication is managed by the local process only. The remote process is not involved. This has the advantage that there is no inherent synchronization between the local and the remote process in every communication request. Nevertheless, at some point, the remote process needs the information as to whether the data which has been sent to that process has arrived and is valid.

To this end, GASPI provides so-called weak synchronization primitives which update a notification on the remote side. In order to ensure an ordered delivery of data payload and notifications, the notifications have to be posted to the same queue to which the data payload has been posted.

The notification semantic is complemented with routines which wait for an update of a single or even an entire set of notifications. In order to manage these notifications in a thread-safe manner GASPI provides a thread safe atomic function to reset local notification with a given ID. The atomic function returns the value of the notification before reset.

The notification procedures are one-sided and only involve the local process.

```
let buf[0] and buf[1] be data buffers;
let bnd[0] and bnd[1] be boundary buffers;
let hal[0] and hal[1] be halo buffers;
set b to 0;

while (!done)
{
   // send boundary data
   write bnd[b] to remote hal[b]
   // send boundary domain
   // validity notification
   notify remote hal exchange

  // update inner domain
  buf[1-b] = compute(buf[b],bnd[b])

  // wait for halo data validity
  notify_wait_some
```

```
  {
    // atomically reset notification
    notify_reset
    // update boundary domain
    bnd[1-b] = compute(bnd[b],hal[b])
  }
  // swap buffers
  set b to 1-b;
}
```

Listing 2 Computation involving weak synchronization and double buffering

Listing 2 shows a weak synchronization example with double buffering. Here, a non-local operator is propagated in an iteration loop, which uses domain decomposition. This pattern usually appears in a similar way in most parallel stencil algorithms. Due to the non-locality of the computational kernel, the individual GASPI processes are tightly coupled throughout the computation.

The local data is partitioned into three sub-domains. Inner data, boundary data and halo data. Each of these sub-domains is represented by two buffers residing in the local partition of the global address space. They are swapped after each iteration. In order to compute the inner data, information from the boundary data is required. In order to compute the boundary data, the remote boundary data—which is locally stored in the halo data—is required. There hence are data dependencies among the local sub-domains and on an inter-process level. In GASPI, the remote data dependencies are mapped a notifications.

Inside of the iteration loop, first the boundary data is sent to the remote side and a subsequent notification is emitted in order to inform the remote process about the validity of its halo data in the next time step. Subsequently the inner data is updated. The local GASPI process then waits in a notify_wait_some ensuring the halo data validity from the remote side. As soon as the data arrives, the local notification flag is reset and the update of the boundary data is performed. At the end, the data buffers are swapped for the next iteration.

By using the double buffering mechanism together with the weak synchronization, we can avoid all potential data race conditions. The communication of the boundary data completely overlaps with the computation of the inner domain. The weak synchronization reduces the coupling of the processes to a minimum (relaxed synchronization model). The local process is not interrupted by the communication from the remote process.

4.3 Extended Calls

Beside the basic and the weak synchronization calls, GASPI also provides extended one-sided communication calls

- `gaspi_write_notfiy`
- `gaspi_write_list`

- `gaspi_read_list`
- `gaspi_write_list_notify`

These extended calls are semantically equivalent to subsequent calls of basic and/or weak synchronization calls. However, GASPI provides an independent definition in order to allow for implementations which can leverage hardware specific optimizations.

For the write operations, GASPI provides three additional extended calls. One is the communication of a single message combined with a subsequent notification event on the remote side. The remote side can wait on the corresponding notification flag. If the notification event is detected, the data payload has arrived and is valid. The second call initiates a communication call of arbitrarily distributed data in n contiguous blocks. It is semantically equivalent to n subsequent calls of ordinary writes to the same destination rank using the same communication queue.

For this second call a list of offsets for local and the remote side as well as corresponding block sizes has to be provided. The list of offsets does not need to be equal on the local and the remote side. The third call is a combination of the second call with a subsequent notification event on the remote side.

5 GASPI Passive Communication

GASPI also provides two-sided semantics—where a send request requires a matching receiver—in the form of passive communication. Passive communication aims at communication patterns where the sender is unknown (i.e. it can be any process from the receiver perspective) but there is a potentially need for synchronization between processes. Typical example uses cases are:

- Distributed update
- Pass arguments and results
- Atomic operations

The implementation should try to enforce fairness in communication that is, no sender should see its communication request delayed indefinitely.

The passive keyword means that the communication calls avoid busy-waiting, computation. Instead passive receives are triggered by an incoming message in a predefined communication queue.

The send request is asynchronous, whereas the matching receive is *time-based blocking*. Due to the asynchronous nature of the send request, a complete send involves two procedure calls. First, one call which initiates the communication. This call posts a communication request to the underlying network infrastructure. If the application wants to update the data-payload which has been send a second call is required, which waits for the completion of the communication request.

In passing we note that GASPI never uses communication buffers—all calls directly read and write application data.

```
** consumer:
let buffer be one data buffer;
while (!done)
{
  passive_receive into buffer;
  process (buffer);
}

** producer:
let buffer[0], buffer[1] be data buffers;
set b to 0;
while (!done)
{
  produce data in buffer[b];

  wait for the completion of earlier
    passive_send;
  passive_send data from buffer[b]
    to consumer;

  set b to 1-b;
}
```

Listing 3 Single consumer and multiple producers using passive communication. The producer transfers a data packet while producing the next data packet, thus overlapping computation and communication

Listing 3 shows a single consumer multiple producer example using passive communication with overlap of communication and computation on the producer side. The consumer sleeps in a passive receive and waits until some data is received into its receive buffer. As soon as the data has been received, the consumer wakes up and processes the data. This is repeated until some termination condition is met. The producers use double buffering. One buffer is used for the computation. The second buffer is used for the communication. The producer is computing some data along the data transfer. Before the data is sent to the consumer after the computation has been done, a wait operation on the previous non-blocking send is invoked in order to ensure that the communication buffer can be used for storing the result of the next computation. If this is the case, the data transfer is initiated and the communication and computation buffers are swapped. On the producer side, this pattern is also repeated until some termination condition is met.

Such single consumer and multiple producer patterns can be used for example in order to implement a global convergence check which is not time critical and which hence can be offloaded to an extra thread without polluting the actual computation threads.

6 GASPI Global Atomics

GASPI provides atomic counters, i.e. globally accessible integral types that can be manipulated through atomic procedures. These atomic procedures are guaranteed to execute from start to end without fear of preemption causing corruption. GASPI provides two basic operations on atomic counters: `fetch_and_add` and `compare_and_swap`. The counters can be used as global shared variables used to synchronize processes or events. Atomic counters are predestined for the implementation of dynamic load balancing schemes. An example is shown in Listing 4. Here, clients atomically fetch a working package ID and increment the value of the package ID counter by one. In accordance with the package ID a corresponding part of the total work load is performed by the local process. This procedure is repeated until all working packages have been processed.

The standard guarantees fairness, i.e. no process should have its atomic operation delayed indefinitely.

The number of atomic counters available can be defined by the user through the configuration structure at start-up and cannot be changed during run-time. The maximum number of available atomic counters is implementation dependent.

```
do
{
  packet := fetch_and_add (1);
  // increment the value by one,
  // return the old value

  if (packet < packet_max):
    process (packet);
}
while (packet < packet_max);
```

Listing 4 Dynamic work distribution: Clients atomically fetch a packet id and increment the value

7 GASPI Collective Communication

Collective operations are operations which involve a whole set of GASPI processes. That means that collective operations are collective with respect to a group of ranks. They are also exclusive per group, i.e. only one collective operation of a specific type can run at a time. For example, two allreduce operations for one group can not run at the same time; however, an allreduce and a barrier can run at the same time.

Collective operations have their own queue and hence typically will be synchronized independently from the operations on other queues (separation of concerns).

Collective operations can be either synchronous or asynchronous. Synchronous implies that progress is achieved only as long as the application is inside of the call. The call itself, however, may be interrupted by a timeout. The operation is then continued in the next call of the procedure. This implies that a collective operation may involve several procedure calls until completion.

Please note that collective operations can internally also be handled asynchronously, i.e. with progress being achieved outside of the call.

GASPI does not regulate whether individual collective operations should internally being handled synchronously or asynchronously, however: GASPI aims at an efficient, low-overhead programming model. If asynchronous operation is supported, it should leverage external network-resources, rather than consuming CPU cycles.

Beside barriers and reductions with predefined operations, reductions with user defined operations are also supported via callback functions.

Not every collective operation will be implementable in an asynchronous fashion—for example if a user-defined callback function is used within a global reduction. Progress in this case can only be achieved inside of the call. Especially for large systems this implies that a collective potentially has to be called a substantial number of times in order to complete—especially if used in combination with GASPI_TEST. In this combination the called collective immediately returns (after completing an atomic portion of local work) and never waits for messages from remote processes. A corresponding code fragment in this case would assume the form:

```
while ( GASPI_allreduce_user (
          buffer_send
        , buffer_receive
        , char num
        , size_element
        , reduce_operation
        , reduce_state
        , group
        , GASPI_TEST ) != GASPI_SUCCESS )
{
    work();
}
```

Listing 5 Collective communication

Listing 5 shows an user defined allreduce using GASPI_TEST in order to overlap the reduction operation with work.

8 GASPI Failure Tolerance

8.1 GASPI Timeouts

On systems with large number of processes, all non-local communication should be prepared for a potential failure of one of the communication partners. In GASPI this is accomplished by providing a timeout value as an argument to all non-local communication calls and the possibility to check for the state of each of the communication partners.

The timeout for a given procedure is specified in milliseconds. `GASPI_BLOCK` is a special predefined timeout value which blocks the invoked procedure until the procedure is completed. This special value should not be used in a failure tolerant program, because in a situation in which the procedure cannot complete due to a failure on a remote process, the procedure will not return at all.

`GASPI_TEST` is special predefined timeout value which represents a timeout equal to zero. Timeout equal to zero means that the invoked procedure processes an atomic portion of the work and returns after that work has finished. It does not mean that the invoked procedure is doing nothing. It does not mean that the invoked procedure returns immediately.

8.2 GASPI Error Vector

GASPI provides a predefined vector type to describe the state of the communication partners of each local GASPI process. Each entry of this error vector corresponds to one remote process. The state of a remote process is denoted to be either *HEALTHY* or *CORRUPT*. After a failed completion of a non-local procedure call, the error vector can be updated by means of a dedicated update procedure. Subsequently the state of the communication partners can be analyzed. In case of a failure of the communication partners, the process can enter a recovery phase.

Conclusion
We have presented the Global Address Space Programming Interface (GASPI) as an alternative to the programming model of MPI. GASPI is a Partitioned Global Address Space (PGAS) API, targeting both extreme scalability and failure-tolerance. GASPI follows a single program multiple data (SPMD) approach and offers a small, yet powerful API composed of synchronization primitives, synchronous and asynchronous collectives, fine-grained control over one-sided read and write communication primitives,

(continued)

global atomics, passive receives, communication groups and communication queues.

The GASPI standard is currently being implemented in two flavors: a highly portable open source implementation and a commercial implementation aimed at ultimate performance. This latter implementation will be based on GPI. The TU Dresden, ZIH will provide profiling support for GASPI by means of extending the VAMPIR tool suite.

There are a number of related projects which pursue similar goals as GASPI (such as OpenSHMEM). The GASPI hence sees itself as one of the precursors towards establishing a de-facto standard for an API for scalable, fault-tolerant and flexible communication in a Partitioned Global Address Space.

Acknowledgements The authors would like to thank the German Ministry of Education and Science for funding the GASPI project (funding code 01IH11007A) within the program "ICT 2020 - research for innovation". Further more, the authors are grateful to all of project partners for having fruitful and constructive discussions.

References

1. MPI Forum: MPI: A Message Passing Interface Standard, http://www.mpi-forum.org/docs/docs.html (2014)
2. Chamberlain, B.L., Callahan, D., Zima, H.P.: Parallel programmability and the chapel language. Int. J. High Perform. Comput. Appl. **21**(3), 291–312 (2007)
3. Consortium, U.: UPC Specifications v1.2. Lawrence Berkeley National Lab Tech Report LBNL-59208 (2005)
4. Hilfinger, P., Bonachea,D., Datta, K., Gay, D., Graham, S., Liblit, B., Pike, G., Su, J., Yelick, K.: Titanium Language Reference Manual. U.C. Berkeley Tech Report UCB/EECS-2005-15 (2005)
5. Numrich, R.W., Reid, J.: Co-array Fortran for parallel programming. ACM Fortran Forum **17**, 1–31 (1998)
6. Machado, R., Lojewski, C.: The Fraunhofer virtual machine: a communication library and runtime system based on the RDMA model. Comput. Sci. Res. Dev. **23**, 125–132 (2009)
7. Fraunhofer ITWM: GPI - Global Address space programming Interface. http://www.gpi-site.com (2013)
8. Simmendinger, C., Jägersküpper, J., Machado, R., Lojewski, C.: A PGAS-based implementation for the unstructured CFD solver TAU. In: 5th Conference on Partitioned Global Address Space Programming Models, Tremont House, Galveston Island (2011)
9. Machado, R., Lojewski, C., Abreu, S., Pfreundt, F.-J.: Unbalanced tree search on a manycore system using the GPI programming model. Comput. Sci. Res. Dev. **26**(3–4), 229–236 (2011)
10. Poole, S.W., Hernandez, O., Kuehn, J.A., Shipman, G.M., Curtis, A., Feind, K.: OpenSHMEM - toward a unified RMA model. In: Encyclopedia of Parallel Computing, pp. 1379–1391 (2011)

Characteristic Analysis of Applications for Designing a Future HPC System

Osamu Watanabe, Takashi Soga, Youichi Shimomura, and Akihiro Musa

Abstract Computer simulations have become an effective tool for finding solutions to various social and scientific challenges: countermeasures against disasters, environmental issues, industrial competitiveness, and so on. To research future HPC systems for finding such solutions, the Ministry of Education, Culture, Sports, Science and Technology in Japan conducted a research program, "Feasibility Study of Future HPCI Systems". Tohoku University, the Japan Agency for Marine-Earth Science and Technology, and NEC Corporation participated in the program. An application research group was formed to clarify characteristics of several application programs in the fields of natural disaster mitigation and high productivity engineering for designing a future HPC system by around 2020. In this article, we describe the overview of our research in the program.

The application research group investigates characteristics of the application programs; for example, the ratio of memory bandwidth to computational performance (B/F), calculation amount, memory capacity, MPI data traffic, and so on. Then, we clarify that the B/Fs of most researched application programs are greater than 2 B/F. Thus, these application programs are memory intensive, and the B/F in future HPC systems will need 2 B/F or greater to preserve the computational performance. Also, this group estimates the performance of the application programs on our designed future HPC system with a high memory bandwidth. Our research shows that our future HPC system has the potential to overcome the challenges of natural disaster mitigation and high productivity engineering.

O. Watanabe (✉) • Akihiro Musa
1st Government and Public Solutions Division, NEC Corporation, 7-1 Shiba 5-chome, Minato-ku, Tokyo 108-8001, Japan
e-mail: o-watanabe@az.jp.nec.com; a-musa@bq.jp.nec.com

T. Soga
NEC Solution Innovators, 4-24 Shiromi 1-chome, Chuo-ku, Osaka 540-8551, Japan
e-mail: soga-txa@necst.nec.co.jp

Y. Shimomura
NEC Solution Innovators, 10-23 Ichibancho 1-chome, Aoba-ku, Sendai, Japan
e-mail: y-shimomura@wx.jp.nec.com

M.M. Resch et al. (eds.), *Sustained Simulation Performance 2014*,
DOI 10.1007/978-3-319-10626-7_3

1 Introduction

Computer simulations of natural phenomena and engineered systems have become a fundamental tool for studying science and technology and have recently been utilized for solving various social and scientific challenges: countermeasures against disasters, environmental issues, energy issues, industrial competitiveness, and so on. To solve future issues in these fields, the computer simulations will require much higher performance HPC systems. The Ministry of Education, Culture, Sports, Science and Technology conducted a research program ("Feasibility Study of Future HPCI Systems") to research future HPC systems for solving the above social and scientific challenges. Tohoku University, the Japan Agency for Marine-Earth Science and Technology (JAMSTEC), and NEC Corporation participated in the program with the objective of studying a future HPC system for memory-intensive applications, and an application research group was formed to investigate the social and scientific challenges presented by natural disaster mitigation and high productivity engineering by around 2020. This group aims to

- clarify the social and scientific challenges of natural disaster mitigation and high productivity engineering by around 2020,
- select several application programs for resolving the social and scientific challenges, hereinafter referred to as "target applications",
- clarify characteristics of the target applications for designing a future HPC system, and
- estimate performance of the target applications on our designed future HPC system.

Here, in the field of natural disaster mitigation, we focus on weather, earthquake, and tsunami phenomena. This is because Japan often suffers from torrential rains and tornadoes attributed to extraordinary weather. Moreover, future strong earthquakes and high tsunamis such as the 2011 Great East Japan Earthquake are predicted. The natural disasters caused by both earthquakes and tsunamis are complex, so coupling simulations that consist of earthquake and tsunami simulations are widely expected to predict the damage. Meanwhile, in the field of the high productivity engineering, we focus on power-generating technology and aircraft design. This is because increasing power generation efficiency contributes to the prevention of global warming, and the Japanese aircraft industry is expected to grow through developing a next generation regional jet, the Mitsubishi Regional Jet (MRJ) [1].

In this article, we present an overview of our study. The rest of this article is organized as follows. Section 2 describes the social and scientific challenges of natural disaster mitigation and high productivity engineering on which we focus and shows our target applications for overcoming these challenges. Section 3 presents the characteristics of the target applications for designing our future HPC system. In Sect. 4, the performances of the target applications are estimated on our designed future HPC system. Section 5 describes the potential of overcoming the challenges using our designed system. Finally, Sect. 6 summarizes our work.

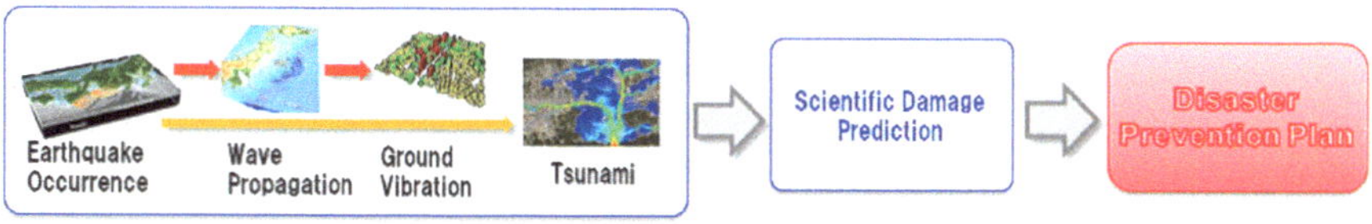

Fig. 1 Coupling simulation for scientific disaster prevention

2 Social and Scientific Challenges

2.1 Natural Disaster Mitigation

Japan often suffers from various natural disasters such as typhoons, earthquakes, and tsunamis. Damage from unexpected torrential rains and tornadoes has been on the increase, and a massive unexpected earthquake occurred on March 11th 2011. The conventional natural disaster prevention and reduction plans have been prepared on the basis of the damage estimated through decades-long experience in Japan. Hence, the natural disasters that occurred during that period were unexpected and sometimes caused considerable damage. Thus, the conventional natural disaster prevention and reduction plans must be reviewed to eliminate the unexpected disasters. Currently, Japan demands that damage be predicted by using scientific approaches with error evaluations for reviewing the convectional natural disaster prevention and reduction plans. Thus, the utilization of computer simulations is expected to predict the damage correctly because the computer simulations can reproduce various natural phenomena in a wide variety of conditions. Furthermore, the coupling simulation of several phenomena can estimate the damage of the complex disaster. Figure 1 shows a coupling simulation of earthquakes and tsunamis. The coupling simulation will be executed using over 1,000 scenarios, which involve various physical conditions of the earthquake and tsunami phenomena in areas of Japan predicted to experience massive earthquakes. Then the 1,000 scenarios will enable damage to be predicted scientifically with error evaluations.

The application programs as shown in Table 1 are our target applications to overcome the challenges of natural disaster mitigation. The coupling simulation of earthquakes and tsunamis consists of RSGDX, Seism3D [2], MMA [3], and STOC-CADMAS [4].

2.2 High Productivity Engineering

Japanese industrial products have been subjected to stiff market competition due to the rapid industrial development of emerging countries. Thus, the international competitiveness of Japanese industrial products needs to be increased for Japan to retain market leadership. In particular, industrial products now need not only

Table 1 Target applications for natural disaster mitigation

Area	Name	Description/method
Earthquake	RSGDX	Seismic cycle simulation
	Seism3D	Seismic wavefield simulation
	QSFDM-GLOBE	Global seismic wave propagation computation
	MMA	Seismic ground response analysis
Tsunami	STOC-CADMAS	Tsunami inundation simulation
	TUNAMI-CADMAS	Tsunami inundation simulation
Climate/weather	CDA [5]	Atmosphere-ocean coupled data assimilation model
	GSM [6]	Global spectral model
	MSSG [7]	Multi-scale atmosphere-ocean simulation

Table 2 Target applications for high productivity engineering

Area	Name	Description/method
Power-generating technology	Numerical turbine [8]	Turbo machinery simulation
Aircraft design	BCM [9]	Aero dynamics simulation
	BCM-LEE	Aero acoustics simulation
CFD	LGAsxmk54v	Lattice gas automate method

high capabilities and low prices but also high safety, high reliability, and low environmental load. Therefore, many Japanese industries have been utilizing computer simulations for research and development of their products and so have achieved design optimizations, short development periods, reliability improvements, and low prices. Hence, the functional enhancements and speeding up of computer simulations are essential in high productivity engineering.

In the field of power-generating technology, the proportion of Japan's energy supplied by thermal electric power systems reached 90 % after the accident at the Fukushima Daiichi nuclear power plant in March 2011. Increasing carbon dioxide emissions has become a serious climate change problem. One solution to it is to develop highly efficient turbo-machinery. Furthermore, fault-tolerant turbo-machinery is required for a steady supply of power. Meanwhile, in the field of aircraft design, the MRJ, which is the nation's first jetliner, will make its maiden flight in 2015. In the future, the Japanese aircraft industry will need to improve the design and development capacity for high safety, high reliability, low environmental load, and low prices.

To overcome the challenges of high productivity engineering, high-resolution simulations are required for all the machinery. Moreover, simulations using a number of different design parameters need to be executed in a short period of time. Here, we select the target applications listed in Table 2.

3 Application Requirements for the Future System

As described in Sect. 2, we treat 13 applications as our target applications for natural disaster mitigation in Table 1 and high productivity engineering in Table 2. Tables 3 and 4 list the simulation sizes (domain size, grid size, simulation time) of the current and future simulations of the target applications, respectively. The domain size and simulation time of the future simulations are almost equal to those of the current simulations, but the resolution of the future simulation size is about 10–20 times finer than that of the current simulation size. This high resolution is necessary to enhance accuracy of the simulations. The high resolution necessitates enormous memory capacity. Therefore, the memory capacities in the future simulations increase 1,000–8,000 times roughly in comparison with the current simulations. Moreover, in general, the finer the resolution of a simulation, the shorter the time interval of time integration due to the Courant-Friedrichs-Lewy (CFL) Condition. Hence the calculation amounts in the future simulations increase 10,000–160,000 times roughly in comparison with the current simulation. Table 5 summarizes requirements of our target application programs for the future HPC system. These requirements are needed to achieve solutions to the above social and scientific challenges. Most of these application programs require over 1 Exa-Flop of the calculation amount. Moreover, some applications need a Peta-Byte class for the memory capacity and storage space. For example, BCM-LEE requires the calculation amount of about 35 Zetta-Flop, Seism3D requires the memory capacity of about 3 Peta-Byte, and MMA requires the disk capacity of about 16 Peta-Byte.

We investigate memory performance of our target applications. Many application programs in HPC generally need high memory performance systems [10]. Figure 2

Table 3 Current simulation size

Applications	Domain size	Grid size/resolution	Simulation time
RSGDX	1,000 km × 300 km	10^5	–
Seism3D	1,200 km × 1,000 km × 200 km	500 m × 500 m × 250 m	600 s
QSFDM_GLOBE	0 ∼ 6,371 km	32×10^6	6,000 s
MMA	100 km × 100 km × 100 km	2×10^8	300 s
STOC-CADMAS	1,400 km × 1,100 km × 10 km	1 m × 1 m × 1 m	3,600 s
TUNAMI-CADMAS	473 km × 724 km × 40 m	1 m × 1 m × 1 m	3,600 s
MSSG	4,000 km × 4,000 km × 40 km	2 km × 2 km × 800 m	3 days
CDA	Global	T42L24	3 months
		1degL45	
GSM	Global	TL319L60	84 h
Numerical turbine	–	91 × 91 × 91	6.7 ms
		13 blocks	
BCM	–	1.7×10^7	100 steps
BCM-LEE	–	1.9×10^7	100 steps
LGAsxmk54v	–	129×10^8	3×10^5 steps

Table 4 Simulation size for ultra-high resolution simulation around 2020

Applications	Domain size	Grid size/resolution	Simulation time
RSGDX	1,000 km × 300 km	10^8	–
Seism3D	1,200 km × 1,000 km × 200 km	40 m × 40 m × 20 m	600 s
QSFDM_GLOBE	0 ∼ 6,371 km	29×10^8	6,000 s
MMA	100 km × 100 km × 100 km	10^{11}	300 s
STOC-CADMAS	1,400 km × 1,100 km × 10 km	0.25 m × 0.25 m × 0.25 m	3,600 s
TUNAMI-CADMAS	593 km × 661 km × 50 m	1 m × 1 m × 1 m	6 h
MSSG	4,000 km × 4,000 km × 40 km	100 m × 100 m × 800 m	3 days
CDA	Global	T216L150	3 months
		1/10degL150	
GSM	Global	TL3839L200	84 h
Numerical turbine	–	91 × 91 × 91	0.017 s
		2813 blocks	
BCM	–	3×10^{10}	10^5 steps
BCM-LEE	–	5×10^{10}	3×10^5 steps
LGAsxmk54v	–	8.8×10^{12}	4×10^6 steps

Table 5 Requirements of target applications for ultra-high resolution simulation

Applications	Calculation (10^{18})	Memory	Disk capacity	Required execution time
RSGDX	1.7	33 TB	26 TB	24 h
Seism3D	870	2.8 PB	433 TB	8 h
QSFDM_GLOBE	86.5	1.7 TB	284 GB	12 h
MMA	1,376	703 TB	16.2 PB	10 days
STOC-CADMAS	116	42.3 TB	4.1 PB	24 h
TUNAMI-CADMAS	19	0.6 TB	378 TB	15 h
MSSG	720	175 TB	1.6 PB	6 h
CDA	2	18 TB	9 TB	4 h
GSM	0.4	21 TB	4 TB	1.4 h
Numerical Turbine (×20)	259	163.5 TB	166 TB	20 h
BCM	33	390 TB	338 GB	0.5 h
BCM-LEE (×10)	35,000	688 TB	72 TB	60 h
LGAsxmk54v	3,300	880 TB	23 MB	24 h

shows the B/F ratio of our target applications. The B/F ratio is the ratio of the memory bandwidth to the peak flop/s, and this ratio is a factor of memory access intensity. The B/F ratio is an important factor to obtain the required performance of the target application programs, so a HPC system needs a large B/F ratio. As shown in Fig. 2, the B/F ratios are about 4.8 on average, although RSGDX, STOC-CADMAS, and TUNAMI-CADMAS have high B/F ratios of 8 or 9. Therefore, the application programs in natural disaster mitigation and high productivity engineering need a high memory performance system.

In natural disaster mitigation, we investigate the coupling simulation as described in Sect. 2. In this investigation of the coupling simulation, the domain area is

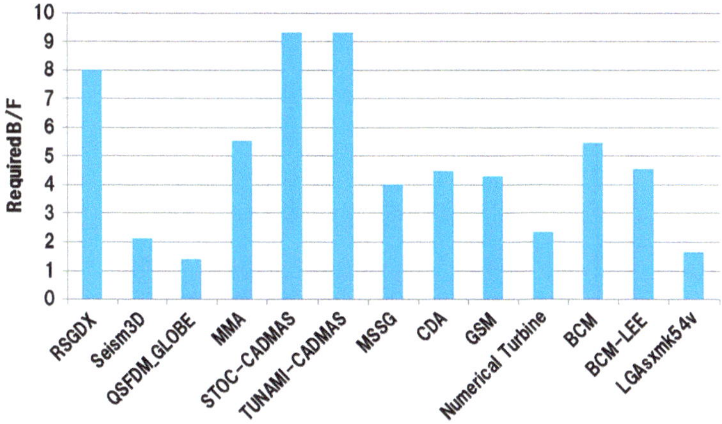

Fig. 2 Required B/F ratio

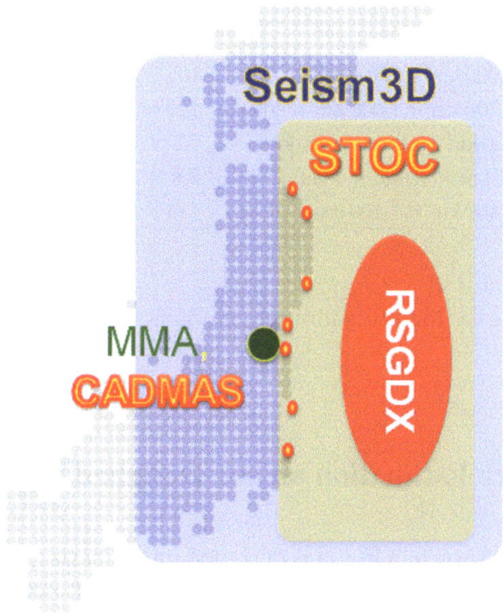

Fig. 3 Simulation domain for disaster mitigation

the Tohoku area affected by the 2011 Great East Japan Earthquake as shown in Fig. 3. Figure 4 shows the simulation flow of the coupling simulation for the disaster mitigation. This simulates earthquake occurrence, tsunami inundation, and strong ground motion. Table 6 indicates the simulation sizes of the application programs constituting the coupling simulation. The required execution time of this coupling simulation is less than 3 h. By this coupling simulation, damage caused by earthquakes and tsunamis can be foreseen within 3 h and the information can

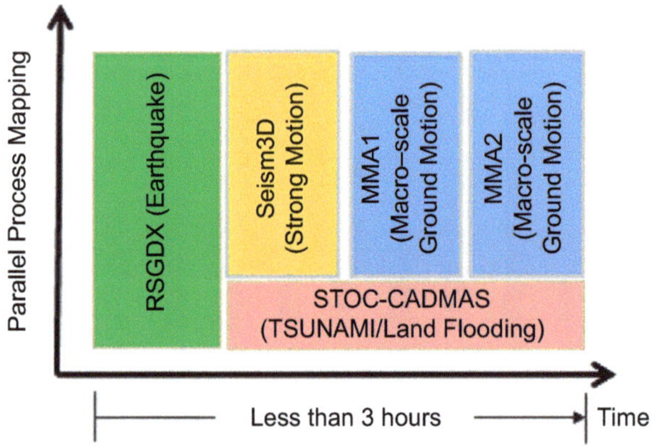

Fig. 4 Execution flow of application programs constituting coupling simulation

Table 6 Simulation size for coupling simulation for scientific disaster prevention around 2020

Applications	Domain size	Grid size/resolution	Simulation time
RSGDX	1,000 km × 300 km	10^7	–
Seism3D	1,200 km × 1,000 km × 200 km	125 m × 125 m × 62.5 m	600 s
MMA1	100 km × 100 km × 10 km	7.8×10^8	300 s
MMA2	10 km × 10 km × 1 km	7.8×10^8	300 s
STOC-CADMAS	1,400 km × 1,100 km × 10 km	1 m × 1 m × 1 m	3,600 s

be utilized as input for the mitigation in emergency situations caused by complex disasters.

4 Performance Estimation on our Designed System

The HPC system as shown in Fig. 5 is designed on the basis of our analysis of these target application programs. Its peak performance is 100-peta flop/s, and its most notable point its high memory bandwidth of a 2 B/F.

For performance estimation of the target application programs, first we analyze the program structure of each program, then estimate the performance of three parts on the target application programs (calculation, communication, and I/O parts), and finally estimate total execution time by summing these three parts. The calculation and communication parts are estimated by using a timing simulator and a network simulator, respectively. The I/O part is estimated by analysis of I/O patterns and overhead of file creation.

To estimate these target application programs, we evaluate them on variations of system specifications: memory bandwidth, network bandwidth, network topology, I/O bandwidth, and so on. Figure 6 shows performance of the coupling simulation

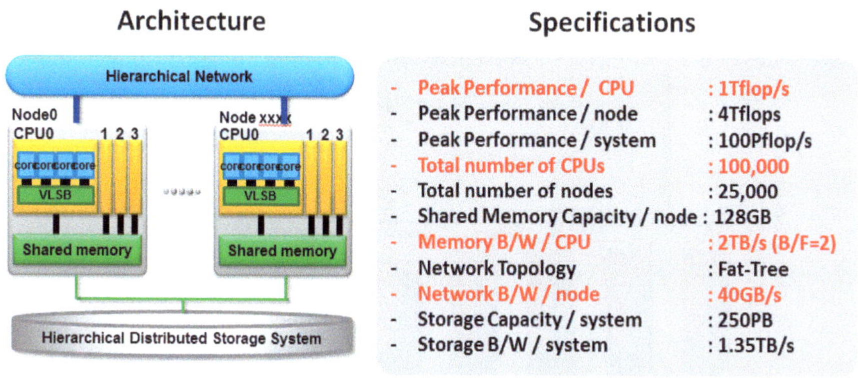

Fig. 5 Target HPC system for estimation

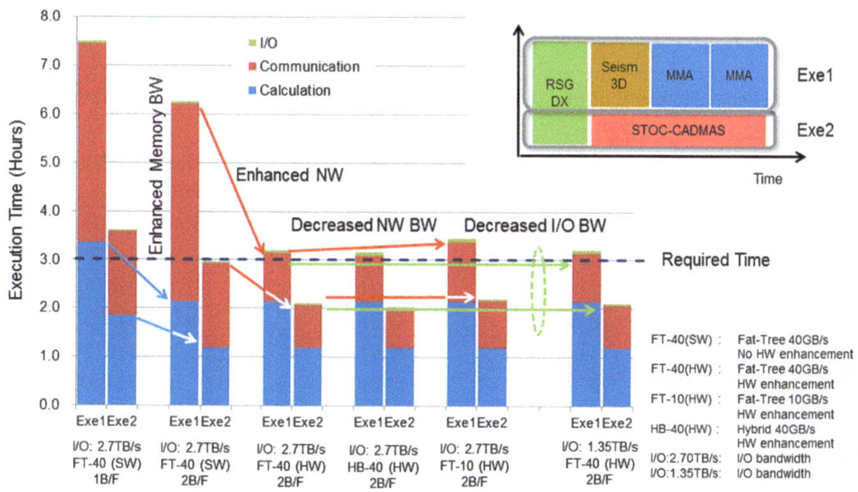

Fig. 6 Performance analysis of coupling simulations for disaster mitigation

for the disaster mitigation under various system conditions. In Fig. 6, the blue, red, and green parts of the bar indicate the calculation, communication, and I/O times, respectively. The leftmost paired bars are the case of 1 B/F memory performance, and the paired bars second from the left are the case of 2 B/F memory performance. Comparing the leftmost paired bars to the right-hand paired bars indicates that the memory performance affects the calculation time effectively. The third paired bars from the left are the case of supporting HW collective communication function. Comparing the second paired bars from the left to the third bars from the left indicates the collective communication affects the communication time dramatically. Regarding this coupling simulation, other factors (network topology, network bandwidth, and I/O bandwidth) have little effect on performance.

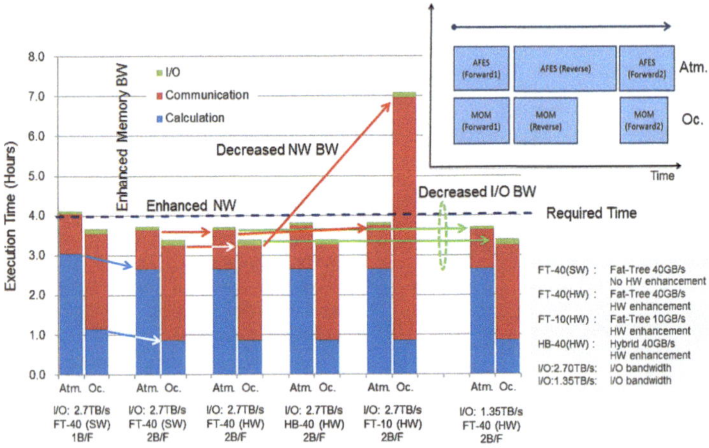

Fig. 7 Performance analysis of CDA

Table 7 Estimated execution time of target application programs in our designed future HPC system

Applications	Required execution time	Estimated execution time	Num of CPUs (Peak Pflop/s)
RSGDX	24 h	4.2 h	100,000 (100 Pflop/s)
Seism3D	8 h	7.1 h	100,000 (100 Pflop/s)
QSFDM_GLOBE	12 h	10.7 h	7,680 (8 Pflop/s)
MMA	10 days	87.2 h	100,000 (100 Pflop/s)
STOC-CADMAS	24 h	22.9 h	98,000 (98 Pflop/s)
TUNAMI-CADMAS	15 h	12.2 h	4,810 (4.8 Pflop/s)
MSSG	6 h	12.6 h	102,400 (102 Pflop/s)
CDA	4 h	3.7 h	8,480 (8.5 Pflop/s)
GSM	1.4 h	1.38 h	76,800 (77 Pflop/s)
Disaster mitigation	3 h	3.22 h	100,000 (100 pflop/s)
Numerical turbine (×20)	20 h	12.9 h	56,260 (56 Pflop/s)
BCM	0.5 h	0.26 h	65,536 (66 Pflop/s)
BCM-LEE (×10)	60 h	51.5 h	100,000 (100 Pflop/s)
LGAsxmk54v	24 h	20.1 h	100,000 (100 Pflop/s)

Figure 7 shows the performance of CDA, which consists of the atmosphere model (AFES) and the ocean model (MOM), under various system conditions. The B/F of the memory system affects the calculation time effectively, especially for the ocean model, and the network bandwidth affects the communication time dramatically, especially for the ocean model.

We estimate performance of our target applications and then accordingly prove that our designed system can achieve the required execution times for most applications as shown in Table 7. Though the calculation time is dominant on many applications, the communication time is dominant on some applications.

In consequence, our designed HPC system is found to satisfy the requirements of almost all the target application programs. The high memory bandwidth and enhanced network performance (HW collective communication function and high network bandwidth) especially contribute largely to achieving this.

5 Potential of Overcoming the Challenges by Using our Designed System

5.1 Natural Disaster Mitigation

The coupling simulation of earthquakes and tsunamis can estimate the damage of a complex disaster. In Sect. 4, each simulation composing the coupling simulation had 1–10 times the resolution of each current simulation. Specifically, the resolution of the tsunami inundation simulation is 1 m, thus the resolution of the simulation is 100 times finer than that of the current disaster prevention and reduction plan. Furthermore, the execution time of the coupling simulation is within 3.2 h per scenario using our designed system. Therefore, the simulation of 1,000 scenarios is executed in 4.5 months. Currently, five Japanese areas (Kuril Trench, Japan Trench, Sagami Trough, Nankai Trough, and Ryukyu Trench) are predicted to experience a massive earthquake and large tsunami. Therefore, the current disaster prevention and reduction plan will be able to be revised on the basis of the simulation of all 1,000 scenarios within two years. Moreover, immediately after a massive earthquake hits, the coupling simulation will predict the damage within 3.2 h. Then the information of the predicted damage will be utilized for conducting rescue and evacuation activities. Meanwhile, the disaster prevention and reduction plan for weather requires a seasonal plan for regional extreme weather. The ultra-high resolution simulation of the weather models is expected to predict probability of regional extreme weather worldwide. In Sect. 4, the resolution of the high resolution weather model was 100 m in urban areas, which is 20 times finer than the resolution of the current model. Moreover, the simulation, which contains 60 scenarios of CDA, 100 scenarios of GSM, and 10 scenarios of MSSG, is executed within three weeks using our design system. Hence, the accurate forecast on the regional extreme weather will make it possible to develop a seasonal disaster prevention and reduction plan.

5.2 High Productivity Engineering

In the Numerical Turbine simulation, current HPC systems are not able to execute a time-dependent analysis of all circumferences due to their very long execution time. In Sect. 4, however, the time-dependent analysis of all circumferences with 20 designed parameters was executed within 14 h using our design system. Here, the execution time provides the possibility for the simulation to analyze multi-physics phenomena: attachment of particles, condensed water vapor, fusion, and corrosion. Hence, the design and development will become refinements, and the products will realize more reliable, lower cost, and lower environment load around 2020. Meanwhile, in the BCM simulation, Sect. 4 showed that the unsteady flow analysis with 30 billion grid points is executed within 30 min using our design system. The experiment efficiency of the BCM simulation is higher than that of a wind tunnel experiment. Therefore, a digital flight simulator, which simulates turbulent airflow and analyzes turning characteristics, will become a reality. In addition, in the BCM-LEE simulation, the aircraft noise analysis on the entire body with 20 designed parameters is executed within 2 days. Hence, the noise reduction design will be realized for environmental compatibility. Therefore, in the field of the high productivity engineering, the innovative design, which has higher reliability, higher efficiency, and lower cost design, will be made possible by around 2020 using the computer simulations.

6 Summary

We investigated the characteristics of 13 application programs in natural disaster mitigation and high productivity engineering for designing a future HPC system by around 2020. To solve various social and scientific challenges, the resolution of the all researched application programs is made finer than that of the current programs. Then, each application program is given 10^4 to 10^5 times the amount of operations of each current application program. Moreover, the B/Fs of the 11 application programs are greater than 2 B/F. Thus, these application programs are memory intensive. Hence, the B/F in future HPC systems will need 2 B/F or greater to preserve the computational performance.

We estimate the performance of the application programs on our designed future HPC system, which has the 100 PFlop/s performance and the 2 B/F memory bandwidth of the memory system. The 12 application programs can satisfy the performances required to overcome the social and scientific challenges. Hence, the information of the predicted damage of a complex disaster will be able to be utilized for conducting rescue and evacuation activities after natural disasters. Moreover, innovative designs will be made possible in high productivity engineering.

In this article, we clarified the social and scientific challenges and the target applications of natural disaster mitigation and high productivity engineering and analyzed the characteristics of the target applications. Finally, we clarified that our designed future HPC system has the potential to overcome the challenges. In the future works, we will optimize the application programs for archiving additional performance on our designed future HPC system.

Table 8 Members of the application research group

Affiliation	Name	
Japan Agency for Marine-Earth	Yoshiyuki KANEDA	Kunihiko WATANABE
Science and Technology	Keiko TAKAHASHI	Yoichi ISHIKAWA
	Shinichiro KIDA	Takane HORI
	Mamoru HYODO	Hide SAKAGUCHI
	Daisuke NISHIURA	
Tohoku University	Satoru YAMAMOTO	Fumihiko IMAMURA
	Shunichi KOSHIMURA	Kenjiro TERADA
	Genti TOYOKUNI	
University of Tokyo	Muneo HORI	Takashi FURUMURA
	Tsuyoshi ICHIMURA	
Japan Meteorological Agency	Hideo TADA	Chiashi MUROI
	Keiichi KATAYAMA	Eiji TOYOTA
	Junichi ISHIDA	Takafumi KANEHAMA
Japan Aerospace Exploration Agency	Kazuhiro NAKAHASHI	
Kanazawa Institute of Technology	Daisuke SASAKI	
Tokyo University of Agriculture and Technology	Shun TAKAHASHI	
Research Organization for Information Science and Technology	Hiroshi MATSUOKA	
Port and Airport Research Institute	Taro ARIKAWA	
NEC Corporation	Akihiro MUSA	Osamu WATANABE

Acknowledgements This work was done in collaboration with our application research group, and many colleagues contributed to it. We particularly thank Dr. Takane Hori and Dr. Kenichi Itakura of the Japan Agency for Marine-Earth Science and Technology, Associate Professor Ryusuke Egawa of Tohoku University, and Yoko Isobe and Akihiro Yamashita of NEC for valuable discussions on this research. We also thank Takashi Abe and Kenta Yamaguchi of NEC Solution Innovators and Midori Kano of IX Knowledge for their support.

Appendix

Table 8 lists the members of the application research group.

References

1. Mitsubishi Regional Jet: Overview of the MRJ. http://www.mrj-japan.com/outline.html (2008)
2. Furumura, T., Chen, L.: Large scale parallel simulation and visualization of 3D seismic wavefield using the earth simulator. Comput. Model. Eng. Sci. **6**(2), 153–168 (2004)
3. Ichimura, T., Fujita, K., Tanaka, S., Hori M., Lalith, M., Shizawa, Y., Kobayashi, H.: Physics-based urban earthquake simulation enhanced by 10.7 BlnDOF × 30 K time-step unstructured

FE non-linear seismic wave simulation. In: Proceedings of the International Conference for High Performance Computing, Networking, Storage and Analysis, New Orleans (2014)
4. Arikawa, T.: Super roller flume for computer aided design of maritime structure, CADMAS-SURF/3D. In: Proceedings of The 34th International Conference on Coastal Engineering, Seoul (2014)
5. Sugiura, N., Awaji, T., Masuda, S., Mochizuki, T., Toyoda, T., Miyama, T., Igarashi, H., Ishikawa, Y.: Development of a four-dimensional variational coupled data assimilation system for enhanced analysis and prediction of seasonal to interannual climate variations. J. Geophys. Res. **113**, C10017 (2008)
6. Japan Meteorological Agency: Outline of the operational numerical weather prediction at the japan meteorological agency. http://www.jma.go.jp/jma/jma-eng/jma-center/nwp/outline2013-nwp/index.htm (2013)
7. Takahashi, K., Onishi, R., Sugiyama, T., Baba, Y., Goto, K., Fuchigami, H.: Seamless simulations in climate variability and HPC. In: Resch, M., et al. (eds.) High Performance Computing on Vector Systems 2009, pp. 199–219. Springer, Berlin (2009)
8. Miyaki, S., Yamamoto, S., Sasao, Y., Momma, K., Miyawaki, T., Ooyama, H.: Unsteady flow effect on nonequilibrium condensation in 3-D low pressure steam turbine stages. In: Proc. ASME Turbo Expo 2013, San Antonio, GT2013-94832 (2013)
9. Nakahashi, K., Kim, L.S.: High-density mesh flow computations by building-cube method. In: Groth, C., Zinggm, D.W. (eds.) Computational Fluid Dynamics 2004, pp. 121–126. Springer, Berlin (2006)
10. Kobayashi, H.: Implication of memory performance in vector-parallel and scalar-parallel HEC. In: Resch, M., et al. (eds.) High Performance Computing on Vector Systems 2006, pp. 21–50. Springer, Berlin (2006)

Enhancing High Performance Computing with Cloud Concepts and Technologies

Bastian Koller and Michael Gienger

Abstract With the advent of Cloud Computing and the hype along this term, developments in Clouds have lead to many mature technologies and solutions which are available up to day. Whilst for a long time Clouds have been seen as a wanna-be competitor to High Performance Computing (HPC) and thus the attitude of HPC-affiliated entities was rather to ignore them, their capabilities currently show a good potential for providing the means for extension of HPC portfolios.
Within this chapter we will present how the different aspects of Clouds can be seen, what are the pros and cons of Cloud technologies and concepts and what kind of models are currently discussed. It will give details about ongoing activities in Europe with respect to the provisioning of HPC Clouds and an initial analysis of needed improvements to further extend the synergistic use of Clouds and HPC.

1 Introduction

Today's market is saturated with Clouds. Society is confronted with the term in a variety of use-scenarios, starting from TV adverts of using Clouds for storing private documents up to the hype around BigData Cloud as well as they are currently also criticized, e.g. in conjunction with the recent NSA scandals.[1] Having said this, it is obvious that there is not a single definition of Cloud (solutions) for everyone and that the different application fields also provide the potential to enhance other well-established technologies with these paradigms.

The aim of this text is to present a short overview of High Performance Computing and Cloud (concepts), concentrating on the advantages and disadvantages of these technologies and presenting how they can complement each other.

[1] David Linthicum: http://www.infoworld.com/d/cloud-computing/thanks-nsa-youre-killing-the-cloud-220434

B. Koller (✉) • M. Gienger
High Performance Computing Centre Stuttgart, Nobelstrasse 19, 70569 Stuttgart, Germany
e-mail: koller@hlrs.de; gienger@hlrs.de

M.M. Resch et al. (eds.), *Sustained Simulation Performance 2014*,
DOI 10.1007/978-3-319-10626-7_4

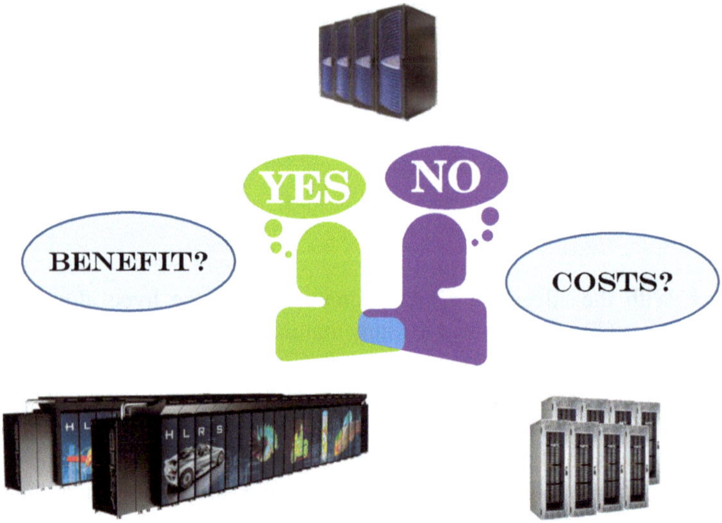

Fig. 1 Costs vs benefits of HPC

2 Current Situation in HPC

Currently High Performance Computing provides large systems which are mainly targeting large industrial customers. Whilst this was in past the best way, the market sees as of today an opening to a wider set of end user communities, moving in a use-model which could be labeled as "HPC-for-everyone".

With this move, new burdens to be overcome appear which need to be addressed properly to realize this adapted business model. The complexity of the access to and use of HPC systems needs to be addressed (and reduced), as well as the means need to be provided to give a clear indication of the benefit-cost relations of using HPC (Fig. 1). The latter applies especially for industrial users, in particular Small and Medium Enterprised (SMEs). Non-HPC-affiliated users (e.g. manufacturing SMEs from new communities) need proper guidance from HPC Experts for the introduction to HPC as a solution. This starts with help in identifying the best-fitting system for the specific purpose of the user. Once (the system is) identified, mechanisms are needed to support the execution of the application whilst supporting ease-of-use for the end user. This is a must as knowledge and experience in HPC is from an economic viewpoint very expensive. Newcomers using first time HPC (systems) simply do not have the experience of using those systems and it is no alternative to try to have all the necessary knowledge in house as this would (a) imply a long learning exercise and (b) be expensive and time consuming without having an indication of potential benefits.

To conclude this: two main questions need to be answered by HPC experts when dealing with new communities using HPC:

- "What does it cost?"
- "What is my benefit?"

These issues and first plans how to overcome them have been addressed e.g. in the PlanetHPC Roadmap[1].

But it has to be clear that answering this question is still easy when being an expert in HPC but for non-experienced users this is rather impossible without any support. Thus its not only about the knowledge in HPC which needs to be provided, also proper mechanisms to access HPC in a variety of ways to gain knowledge need to be delivered to the communities. Ideally this happens in an (at least) semi-automated way, reducing the investments on both sides (customers and provider) and leading to quicker, though probably non perfectly optimized, results. In a lot of ongoing activities, this semi-automation and thus transparency to the end user can be realized on base of Cloud concepts.

3 High Performance Computing and/or Clouds

As previously stated, Clouds and HPC are not necessarily competing with each other. In some areas the one or other concept is the better one to apply and in some cases they can be even used in an complementary way. Therefore it is of utmost importance to understand the capabilities of HPC and Clouds to have a good knowledge base for deciding on what to apply. Therefore the following sections will give a short overview on the comparison of HPC and Clouds and propose models on how to use them in a complementary way.

3.1 High Performance Computing Compared with Clouds

Compared to Cloud, High Performance Computing scales over parallel processes/threads. This improves the overall performance but is dependent on the availability of the underlying infrastructure. In opposite, Clouds are intended to enable scaling over an amount of (service) instances, thus improving the availability, but usually lacking of performance within the provided infrastructure. A potential combination of both, high available and high performance infrastructures would be a clear business benefit for the related communities.

However, these are not the only differences. Several others can be identified, as listed as a short overview below:

- Clouds usually focus less on minimized latency (applications are usually latency tolerant) whilst the HPC applications request low latencies and thus the respective infrastructure.

- Access to Clouds benefits from the main principle of ease-of-use whilst access to HPC is still difficult and requires expertise (not only due to security reasons).
- "Nodes" in HPC are compute units whilst "Nodes" in Clouds can be seen as instances hosting service capabilities.
- Cloud concepts and technologies support a wide range of platforms whilst the range of platforms in HPC is limited.

3.2 *Complementary Use of HPC and Cloud*

Taking into account the aforementioned differences, though partially having complementary capabilities, the question arises how High Performance Computing and Cloud can be combined to provide improved services to its customers and to generate business benefits for all involved stakeholders.

Looking at ongoing activities, there are currently two main trends of Cloud and HPC Combination:

- **The Hybrid Model:** Assumes that Cloud and High Performance infrastructures can be complementary used based on the respective application domains. This includes the concept of splitting workflows in HPC and Cloud parts and executing the different tasks on the respective infrastructures. Those parts being put on Clouds should be latency-tolerant and scalable, those on HPC should be the performance intensive ones. Naturally the management of such workflows on the hybrid model will imply a dedicated and intelligent federation management to realize high level scheduling. Furthermore within the domain of Clouds, job submission from Virtual Machines needs to be realized. First approaches in this direction are taken in a variety of activities (e.g. the DreamCloud project[2]).
- **The Cloud-Like Access Model:** Enhances HPC with the concepts of Clouds in terms of ease-of-use and general simplicity. To hide the complexity of HPC to the end user, automated parts can be provided in a Cloud-like environment. This can happen either as a portal with a simple GUI or, in a more elaborated way, by providing the end users their preferred and automatically set up environment as a Virtual machine. The 2020 Roadmap Consultation document of the European Factories of the Future Research Association (EFFRA)[3] has identified this as an important enhancement of provisioning of HPC and defined the following:

> It is clearly an important priority to address the benefits that SMEs could realize from novel, Cloud-based usage models and to tackle the skills' deficiency in order to enable effective exploitation of the Cloud. What is needed is a one-stop, pay-per-use service where computing, software licensing and domain expertise are bundled together.

Both models propose an added value to current HPC provisionings and even more, they do not exclude each other. It is quite likely that at a certain point in time both models can be used in a combined fashion.

4 Cloud Based Access to HPC: Fortissimo as an Example

Whilst the hybrid model mainly needs solutions on a technical (implementation) level, the Cloud-like access model needs also to take into account the business model(s) of infrastructure providers and the implications of changes, especially in a federated environment. This section will concentrate on the Cloud-like access to HPC and present an ongoing European Commission funded activity, which focuses on offering a marketplace for Small and Medium Enterprises to get simple access to HPC—the Fortissimo project.[2]

4.1 *Introducing the Fortissimo Project*

The Fortissimo (Factories of the Future Resources, Technology, Infrastructure and Services for Simulation and Modelling) Project started on the 1st of July 2013 as an activity within the domain "Factories of the Future", funded by the European Commission within Framework Programme 7. The project is focused on enabling Small and Medium Enterprises to make use of simulations to improve their business capabilities. Therefore it aims to provide a marketplace/one-stop-shop for SMEs providing access to hardware, expertise, software and tools in a considerable way. Within the projects duration (36 months) around 50–60 SME-driven experiments will be executed to refine the requirements to such a one-stop-shop and to refine the design of the offerings. In the first year the project started with 20 experiments, during its evolution two open calls will be performed to increase the number of experiment. As Fortissimo addresses the whole business lifecycle, the consortium involves partners including Manufacturing Companies, Application Developers, Domain Experts, IT Solution Providers and HPC Cloud Service Providers from currently 14 countries (numbers will change with new experiments coming in and others ending).

4.2 *Realizing the One-Stop-Shop*

Fortissimo as a project has as main task to address shortcomings of current HPC offerings with a dedicated focus on easy access and ease of use of those offerings. Therefore a "Cloud of HPC resources" is built which shall provide businesses what they need, e.g. providing them single entry points to heterogeneous special purpose HPC-Systems. Furthermore a dedicated support infrastructure needs to be set up in alignment with what is currently existing at the single centres and extending

[2]The Fortissimo Project—http://www.fortissimo-project.eu/

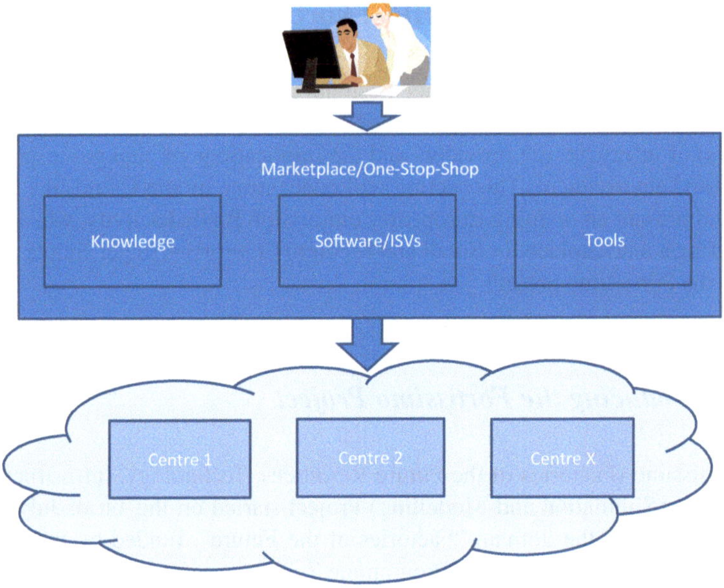

Fig. 2 The Fortissimo concept

this as necessary. New models for licensing (e.g. on-demand licensing) need to be elaborated to ensure proper business models retrieval.

In terms of technology, the aim is to use Cloud and HPC technologies as they are and not to re-invent the wheel or invest massively in the development of new technologies. This also implies to take into account the existing systems on site of the providers (batch systems, etc.) to allow for easy integration of the currently involved but also potential new infrastructure providers.

The initial version of the marketplace (link can be found on the Fortissimo webpage) covers a set of general purpose pages, informing the visitors about infrastructure providers and their capabilities. Furthermore it provides information about available software, expertise of the offering entities and everything needed to get into contact with those (Fig. 2).

The main principle of Fortissimo is to provide simple access to information and to distinguish itself from other HPC/Cloud offers such as Ubercloud[3] or HelixNebula.[4] One distinction is to go beyond just an establishment of contact with the providers (and potentially to set up contracts) where it is then mostly the burden of the end user to find out how to use the system best.

Fortissimo goes beyond this by establishing the means for close interaction of end users and the involved stakeholders in a simulations lifecycle (e.g. code

[3]The Ubercloud—https://www.theubercloud.com/

[4]HelixNebula—http://www.helix-nebula.eu/

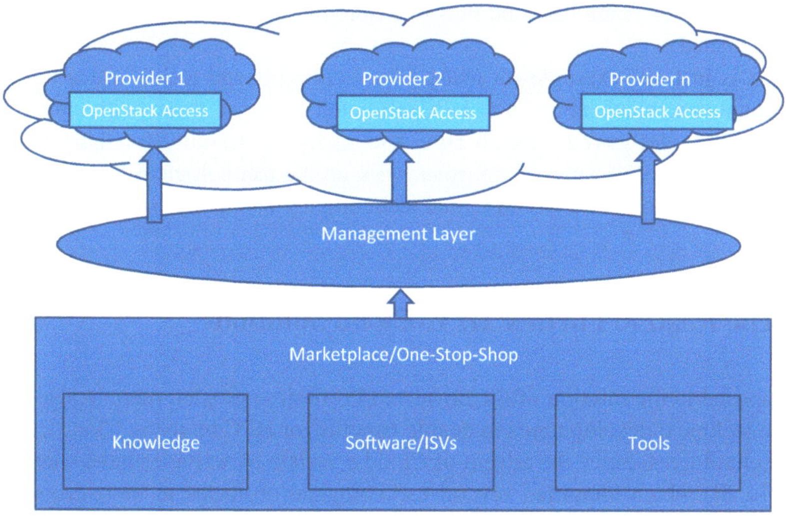

Fig. 3 Job submission via Fortissimo

owners, ISVs) by providing the support infrastructure as well as further automated mechanisms for accessing and using the provided resources. As a prototypical implementation job submission is currently implemented, which is realized in two different ways. On one hand a simple embedding of a provider specific GUI into the marketplace is realizing job submission on some of the Fortissimo centres. On the other hand a more complex solution is implemented and tested, using severe capabilities of the OpenStack[5] Cloud middleware. Figure 3 shows a high level overview of the concepts. The job submission GUI will be situated on the marketplace which is hosted outside of the infrastructure providers domain. To submit the jobs each user gets a virtual machine hosted on an OpenStack environment within the domain of the provider. From this VM, which can act as a virtual user, jobs are submitted internally to the systems, thus overcoming security issues such as opening the firewalls or giving credentials for access to the systems to the marketplace as a third entity. The credentials are stored locally on the providers environments and a mapping between these credentials and the marketplace credentials is stored in a mapping database when the user requests an access to the machines (in a formal way by filling in the required documents and agreeing to the policies of the providers). When the job is executed, status information is pulled by the VM and transmitted to the marketplace for the user.

The user-specific VM provides a good extension potential. In general it enables on the long term to provide the user its personalized environment with help services such as databases when needed or a virtual storage. The latter is especially used for

[5]OpenStack—http://www.openstack.org/

transferring back result data and thus to ensure that the system is cleaned when the job is finished.

This is just one example for using Cloud concepts and technologies to realize functionalities which allow non-HPC affiliated users or those with only minor knowledge to get easy access to HPC and verify its usability for their business. Nonetheless there is a need for further work on the technological side to optimize and enable in general the use of Cloud technologies also in the HPC world.

5 The Road to Further HPC-Cloud Solutions

As opposed to the mission of the presented Fortissimo project, there is also a need on the technical developments to enable benefits for HPC by using Cloud concepts and technologies. Since the advent of Cloud a variety of activities have taken place (e.g. [4, 5]) which identified several technological obstacles to be overcome. On the one hand, reducing the performance losses of a Cloud infrastructure including virtualization compared to a high performance cluster (e.g. [5,6]) or Supercomputer can be handled by upgrading the Cloud hardware to modern high performance standards: fast interconnects like Infiniband or high disk performance with more than 100 GB/s of sustained I/O performance are not very common in current Cloud environments and most of the time, not necessary to provision the Cloud services. On the other hand, the software stack used for handling the virtualization represents the major bottleneck of high performant systems. If full virtualization capabilities are used, hypervisor technologies (like Xen or KVM) foresee the virtualization of one or several additional guest operating systems which result in huge virtualization overheads for the underlying physical system. This paradigm shows that de facto two or more full operating systems are running on a single physical host and disables high performance computing in Cloud infrastructures (Fig. 4).

The design of Cloud environments makes use of one of the biggest advantages of virtualization: flexibility, especially when resources are getting spare and reactions need to happen. Other advantages, such as the encapsulation of services, the reduced costs when using virtualization, the portability of Virtual Machines and the manifold cost and business models provide clear benefits to the users. However, the performance (as mentioned above) usually decreases when using a virtualized system as opposed to a non-virtual system. In parallel security issues arise when using virtualization as this concept bases on the sharing of all hardware resources.

To increase the performance of a virtualized system in order to allow high performance computing, there is the highest potential when changing current hypervisor technology to a more efficient virtualization approach: avoiding the double kernel virtualization whilst at the same time, keeping encapsulation of services. As all modern Linux operating systems are relying on the same management mechanisms like Kernels, System Libraries as well as Kernel Modules, re-using those components and re-combining them between the physical host operating system and the virtualized ones are the key to success. The re-combination of

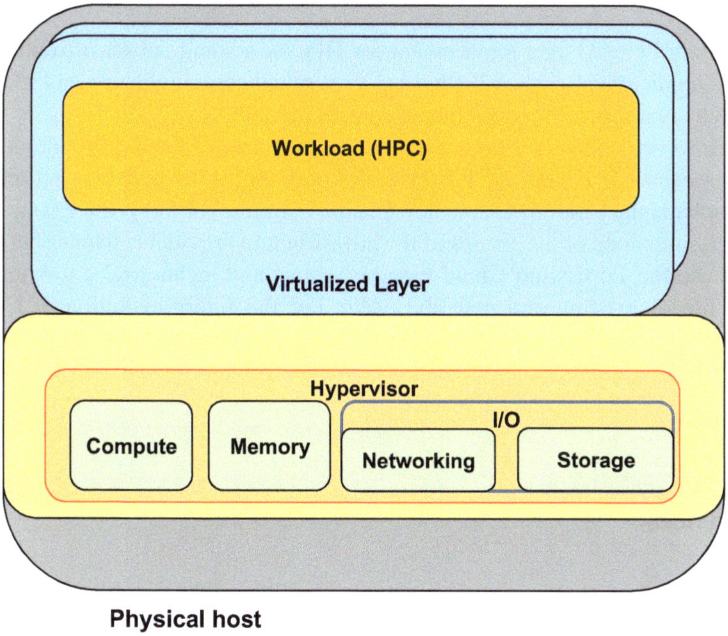

Fig. 4 Sources of overhead in a virtual environment

kernels and modules will result in another performance advantage: kernels and modules can be developed in a small and efficient fashion, functionality can be shared and be separated at the same time. Thus, this approach will not only affect computational power, it will massively improve memory and disk access as well as network performance as all of the mandatory system functions are managed and operated by the underlying physical host.

This new technological approach will not open only paths into flexible high performance Cloud Computing, it will also enrich the Cloud services by faster response times and improved dynamic reactions as well as elastic features. Full virtualization will still be possible, but strong interaction between the components will decrease the virtualization overhead and thus will allow high performance enabled virtualization techniques.

Conclusions

This chapter has provided an overview about High Performance Computing and Cloud Computing and shown that both are not necessarily competing. Both technologies have their advantages and disadvantages and thus relate to certain application areas with a minimal overlap. Clouds, as they are today,

(continued)

can only be used as a replacement for HPC in a small subset of traditional HPC applications, especially those with minimal communication and I/O and which are at the same time embarrassingly parallel.

However, there is a well proven base for using Cloud technologies as the means for extending the HPC services portfolio and being used as complementary tool to enable new features of HPC. In the given example of Fortissimo none of the set ups of the infrastructure providers changes but they can use the Fortissimo Cloud based concepts and technologies to simplify the life of existing and new end users. For the future evolution of Cloud technologies there is certain potential to improve further the performance especially with respect to the virtualization overheads.

References

1. Sawyer, M., Parsons, M. (eds.): Challenges facing HPC and the associated R&D priorities: a roadmap for HPC research in Europe. http://www.planethpc.eu/images/stories/planethpc_roadmap_2013.pdf (2013)
2. Cheptsov, A., Koller, B.: Revisiting dynamic scheduling techniques for HPC infrastructures: the approach of the DreamCloud Project. J. Innov. Supercomput. Dtschl. **11**(2), 50–51 (2013)
3. The European Factories of the Future Research Association (EFFRA) - Multi-annual roadmap for the contractual PPP under Horizon 2020. http://www.effra.eu/attachments/article/129/Factories%20of%20the%20Future%202020%20Roadmap.pdf (2013)
4. Desai, M.: High performance computing and virtualization. Research Paper: CSCi-555 Advanced Operating Systems. http://www-scf.usc.edu/~mitdesai/gradcourse/hpc_virtualization.pdf (2012)
5. Mergen, M.F., Uhlig, V., Krieger, O., Xenidis, J.: Virtualization for high-performance computing. http://anibal.gyte.edu.tr/hebe/AblDrive/69276048/w/Storage/104_2011_1_601_69276048/Downloads/m12.pdf (2011)
6. Gordon, A., Amit, N., Har'El, N., Ben-Yehuda, M., Landau, A., Schuster, A., Tsafrir, D.: ELI: bare-metal performance for I/O virtualization. In: Architectural Support for Programming Languages & Operating Systems, pp. 411–422. ACM, New York (2012)
7. Huber, N., von Quast, M., Hauck, M., Kounev, S.: Evaluating and modeling virtualization performance overhead for cloud environments. CLOSER, pp. 563–573. SciTePress, Beijing (2011)

SX-ACE, Brand-New Vector Supercomputer for Higher Sustained Performance I

Shintaro Momose

Abstract NEC has just launched its new generation SX vector supercomputer, SX-ACE as the successor model of the SX-9 by aiming at much higher sustained performance particularly for memory-intensive applications. This system is based on the high performance core strategy targeting higher sustained performance. It provides both the world top-class single core performance of 64 GFlop/s and the world highest memory bandwidth per core of 64 GBytes/s. Four cores, memory controllers, and a network controller are integrated into a processor as a single LSI, enabling the processor performance of 256 GFlop/s and the memory bandwidth of 256 GBytes/s. Moreover, in order to boost the performance of memory-intensive applications, each core is designed to utilize the whole memory bandwidth of the processor (256 GBytes/s). This brand-new system is quite beneficial to real scientific and engineering applications.

1 Introduction

NEC has always pursued high sustained performance in scientific applications with the SX Series vector supercomputer during the past three decades. As shown in Fig. 1, its first-generational system SX-2 was released in 1983 as a world first system with a peak performance exceeding 1 GFlop/s (GF). After the launch of the pioneering models, NECs HPC solutions have been represented by the unrelenting evolution of this series toward the recent model SX-ACE (see Fig. 2) as the successor to the SX-9 [1–4] with innovative hardware and software technologies.

There are three major design concepts for the SX-ACE system. First, SX-ACE is designed to provide higher sustained performance in real scientific applications same as the previous models of the SX Series. In order to maintain this concept, the SX Series has continuously provided the worlds top-class single core performance with the largest memory bandwidth per core. The SX-ACE single core also inherited such a big core strategy. Its single core performance is 64 GF with 64 GBytes/s (GB/s) memory bandwidth per core, which can reduce the number of necessary

S. Momose (✉)
NEC Corporation, IT Platform Division, 1-10 Nisshin-cho, Fuchu-shi, Tokyo 183-8501, Japan
e-mail: s-momose@ak.jp.nec.com

M.M. Resch et al. (eds.), *Sustained Simulation Performance 2014*,
DOI 10.1007/978-3-319-10626-7_5

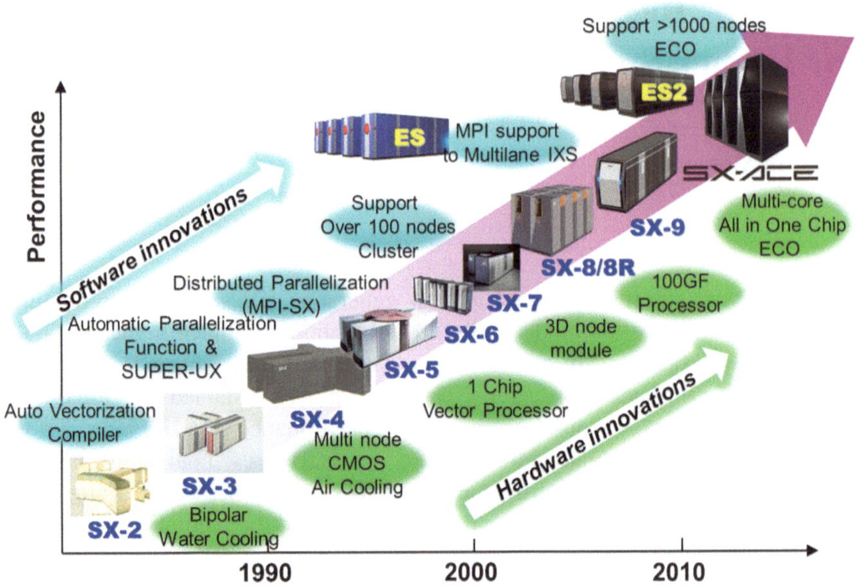

Fig. 1 History of the SX Series and the brand-new vector supercomputer SX-ACE

Fig. 2 SX-ACE system

cores in parallel execution compared to current commodity-based processors characterized by relatively lower performance and lower memory bandwidth per core. The second concept is the reduction of power consumption; SX-ACE realizes smaller power consumption per peak performance than this precedent model SX-9 by a factor of ten. Thirdly, SX-ACE aims at a smaller installation space; it requires only one fifth of the area over the SX-9.

Through the improvement of both hardware and software technologies based on these three design concepts, SX-ACE can significantly contribute to reduced total cost of ownership (TCO).

The rest of this paper describes the architecture of SX-ACE with respect to its processor and system in Sect. 2, and its implementation details in Sect. 3. Section 4 discusses the sustained performance by some benchmark programs, followed by the summary of the paper in section "Conclusions".

2 Architecture of SX-ACE

The SX-ACE processor realizes the world top-class performance and a larger memory bandwidth, both of which inherit the proven SX vector architecture. Figure 3 shows the overview of the processor configuration. It provides a 256 GF double precision floating-point performance and a 256 GB/s memory bandwidth, enabling a ratio of memory bandwidth to performance (Bytes/Flop, B/F) of 1.0. In contrast to B/F ratios lower than 0.5 for current microprocessors, such a high memory bandwidth is effective in enhancing the performance of memory-intensive applications.

The SX-ACE processor is configured with several major blocks, including four processor cores, 16 memory controllers (MCs) that realize a 256 GB/s memory bandwidth, a remote access control unit (RCU), and a crossbar switch. Each core

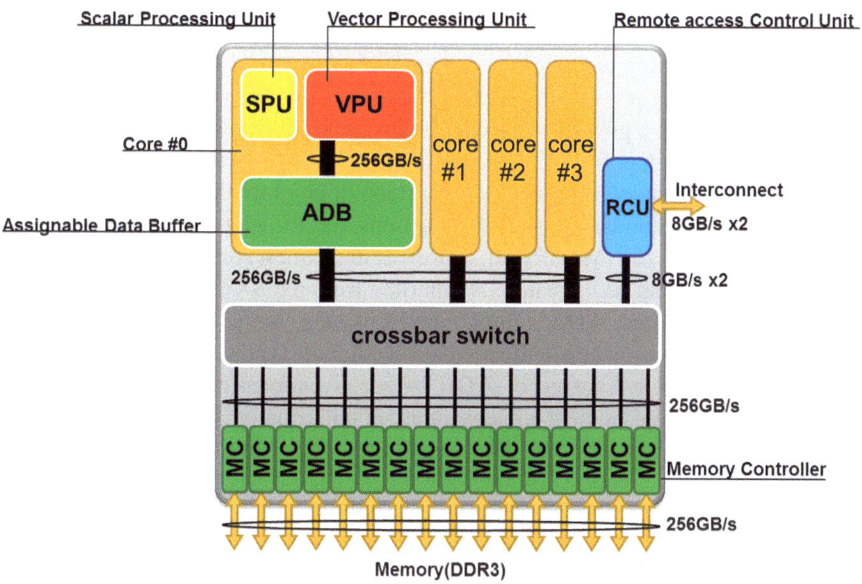

Fig. 3 Processor configuration of SX-ACE

having a performance of 64 GF is composed of three major parts; a scalar processing unit (SPU), a vector processing unit (VPU), and an assignable data buffer (ADB). SPU mainly works as a VPU controller by issuing vector instructions to VPU. VPU is the main calculation part of the core, and provides 64 GF performance. Such an SPU-VPU combined configuration is a major feature inherited from the previous SX vector architecture. ADB has a 1MB capacity with a 256 GB/s bandwidth between ADB and VPU, enabling the B/F ratio of 4.0. MCs have DDR3-based memory interfaces running at 2000 MHz. RCU is a remote direct memory access (RDMA) engine that can be directly connected to a proprietary interconnect network at a transfer rate of 8 GB/s per direction. These components are interconnected through the crossbar switch, which is capable of supplying sufficient data to each component and reducing data transaction conflicts in the data paths. Here, each core can utilize the whole memory bandwidth of 256 GB/s for the processor, as far as the other three cores do not access the memory. This is because the aggregate data transfer rate between the crossbar and the memory is 256 GB/s, while each core and the crossbar are connected with a throughput of 256 GB/s. Thus, the B/F ratio for a single core ranges from 1.0 up to 4.0 at its maximum depending on the status of memory contention, which is quite beneficial in gaining high sustained performance on real scientific applications.

Figure 4 shows the LSI layout of the SX-ACE processor. Four cores are implemented in the upper and lower areas of the LSI. Each side of the LSI accommodates eight MCs with the DDR3 interface. The crossbar switch is implemented in the central part of the LSI, which is surrounded by four cores and 16 MCs. It reciprocally connects the surrounding components at a high transfer rate of 256 GB/s. RCU is implemented in the central upper area, and is connected to the crossbar. This processor is manufactured with the 28 nm process technology. About two billion transistors are integrated into the LSI, with its area of 23.05 mm by 24.75 mm, running at a 1 GHz frequency. Such a relatively low-processor clock cycle contributes to a power reduction in the processor operation.

The major design policy of the SX-ACE processor is the realization of efficient memory access with lower energy consumption. It employs three major techniques in order to avoid redundant memory transactions, i.e., the adoption of ADB and miss status handling register (MSHR) [5], as well as the request compression function for store operations. The ADB feature is inherited from the SX-9 with the quadrupled 1MB per-core capacity. ADB is implemented in each core to reduce the number of memory transactions, enhance the memory bandwidth, and shorten the data access latency. The major advantage of ADB over conventional caches is its software controllable data replacement strategy to avoid pollution on ADB by non-reusable data. The compiler of SX-ACE provides a function of automatic data assignment. MSHR and the store request compression function are newly introduced in order to reduce the number of memory accesses by avoiding redundant memory accesses. These functions marge such redundant memory load and store operations in the processor before issuing either a load or store request from MCs to the memory.

The SX-ACE single processor and 16 DIMMs with the 64 GB memory capacity configure an SX-ACE single node. Figure 5 shows a configuration comparison

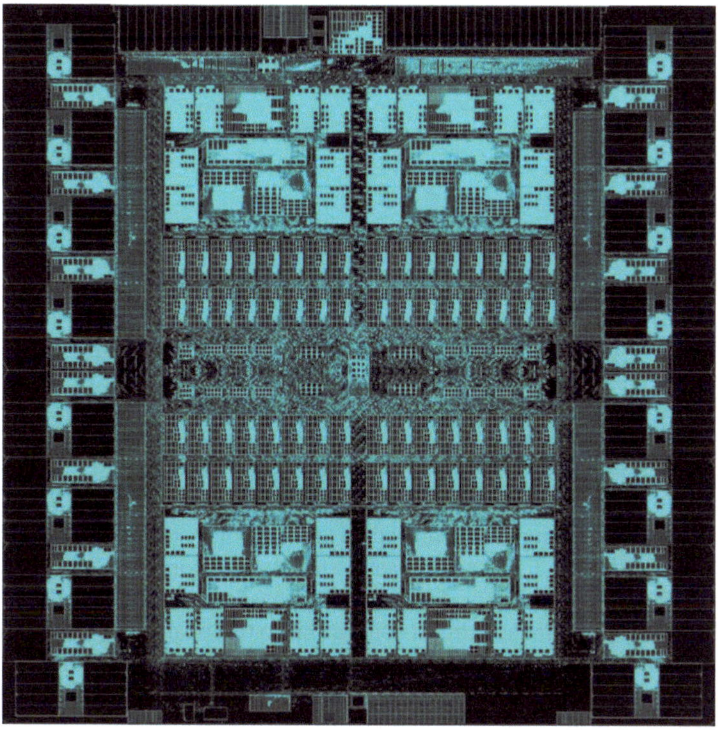

Fig. 4 Processor LSI layout

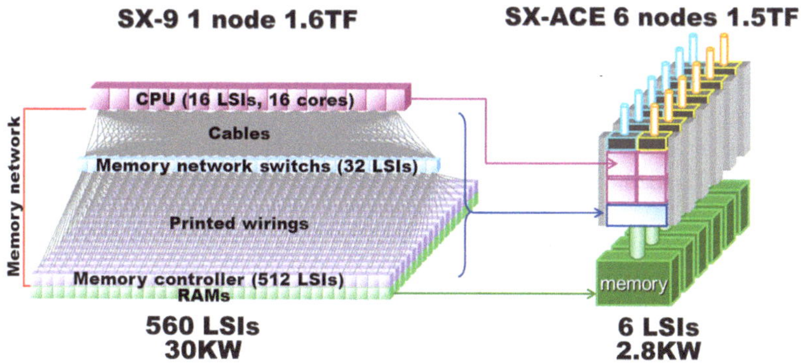

Fig. 5 Comparisons of configuration and power consumption between SX-9 and SX-ACE

between the SX-9 single node and six SX-ACE nodes with a performance of 1.6 and 1.5 TF, respectively. For one node of the SX-9, 16 processors and 1 TB memory are integrated by a memory network as a large SMP configuration with a 4 TB/s memory bandwidth. Each SX-9 node is made up of 560 LSIs with its power consumption of 30 kW. While the SX-9 node has provided higher sustained performance, more than

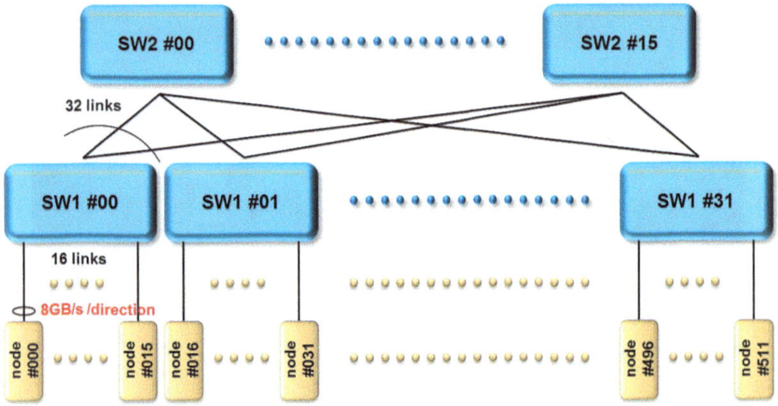

Fig. 6 SX-ACE multi-node system

70 % of its power consumption is attributed to the memory network configured with many LSIs. For SX-ACE, all components, such as four cores, MCs, and RCU, are built into a single processor for reduced power consumption and installation space. Thus, the 1.5 TF performance almost comparable to that of an SX-9 node is achieved by only six SX-ACE processor LSIs. It means the significantly smaller number of LSIs for SX-ACE than the SX-9 by a factor of almost one hundred, resulting in approximately ten times of the reduction in power consumption for SX-ACE in comparison with the SX-9.

As for the SX-ACE multimode system, each node is connected by a proprietary fat-tree topological interconnect network called IXS with an 8 GB/s per-direction throughput as shown in Fig. 6 for a 512-node configuration. Each node is directly connected to a corresponding edge level switch through RCU in each processor. Each switch has 32 ports with a transfer rate of 8 GB/s per direction each and a 32 by 32 crossbar function. In order to accelerate internode communication and barrier synchronization, dedicated hardware functions are implemented into IXS.

3 Implementation

The SX-ACE single node is implemented into a small node card in order to reduce the installation space as shown in Fig. 7. The SX-ACE processor is placed in the central part of the card with eight DIMMs installed on each side. The card size measures 11 cm wide and 37 cm deep.

Figure 8 shows the SX-ACE single rack configuration: two node cards make up one module, eight 2-node modules configure a 16-node cage, and four 16-node cages are implemented into a single rack. Thus, one rack has 64 nodes and its aggregate performance and memory bandwidth reach 16 TF and 16 TB/s, respectively.

Fig. 7 SX-ACE node card

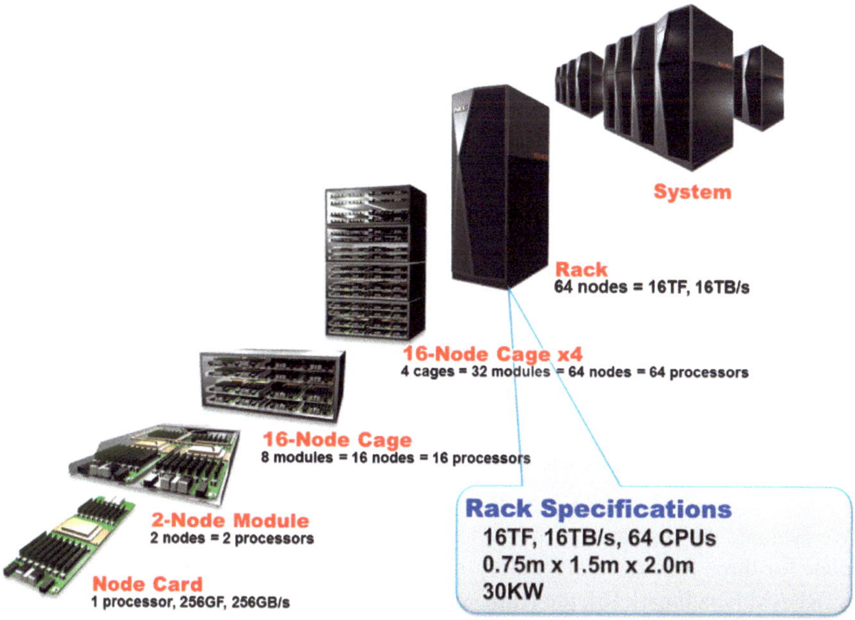

Fig. 8 SX-ACE rack configuration

The rack performance of 16 TF is ten times higher than the SX-9. Moreover, due to the adoption of the big core strategy, such a high performance and a large memory bandwidth are obtained with only 256 cores implemented in one rack. It allows the user to gain higher sustained performance in parallel execution without any significant overheads resulting from massive parallelism with lower performance cores. The rack dimension measures 0.75 m wide, 1.5 m deep, and 2.0 m high, and

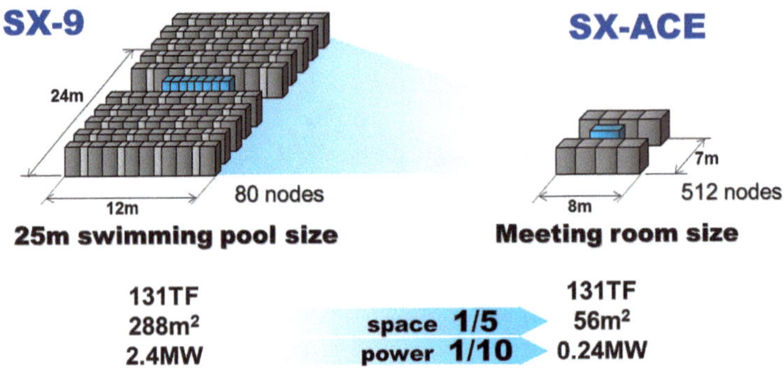

Fig. 9 Installation comparison between SX-9 and SX-ACE

the required power consumption is 30 kW. The hybrid cooling mechanism is adopted for the rack, which combines liquid and air cooling. Processors are cooled by the coolant flow and other components such as DIMMs are cooled by the air flow from the front side of the rack through the back side.

Figure 9 shows the comparison of the SX-9 and SX-ACE with respect to required installation space and power consumption at the same system peak performance of 131 TF. The SX-9 provides a 131 TF performance with 80 racks containing 80 nodes, which requires 288 m^2 of the installation space including the maintenance area, and 2.4 MW power consumption. SX-ACE can provide the same performance with eight racks containing 512 nodes with a much smaller installation space of 56 m^2 and power consumption of 0.24 MW. They are equivalent to only one fifth of the floor space and one tenth of the lower power consumption as compared to the SX-9.

4 Performance Evaluation

We evaluate the sustained performance and power efficiency of the SX-ACE single node for three fundamental memory-intensive benchmark programs, which are the STREAM benchmark [6], the Himeno benchmark [7, 8], and a benchmark program for the Legendre transformation. The STREAM benchmark evaluates the sustained memory bandwidth by disabling data caching functions such as ADB in accessing a data set that has no data locality. The Himeno benchmark is designed to measure the performance in solving the Poisson equation with the Jacobi iterative method, which is highly memory-intensive. This benchmark is used for the evaluation of ADB and MSHR. A separate benchmark program is also prepared to measure the performance of the Legendre transformation often appearing in physics simulation such as the spectral method for weather prediction, and is used for evaluating indirect memory access.

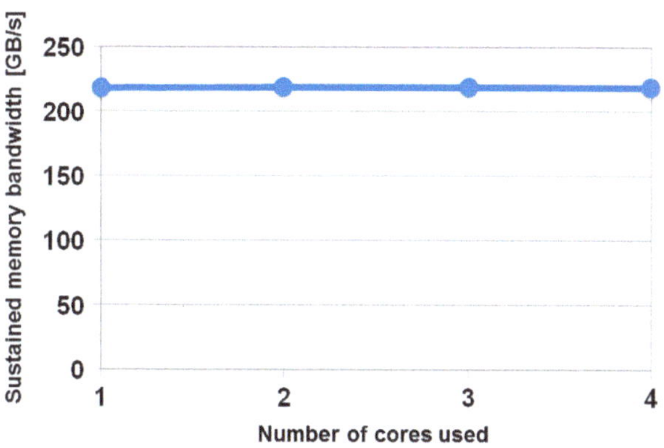

Fig. 10 Performance evaluation results with the STREAM benchmark

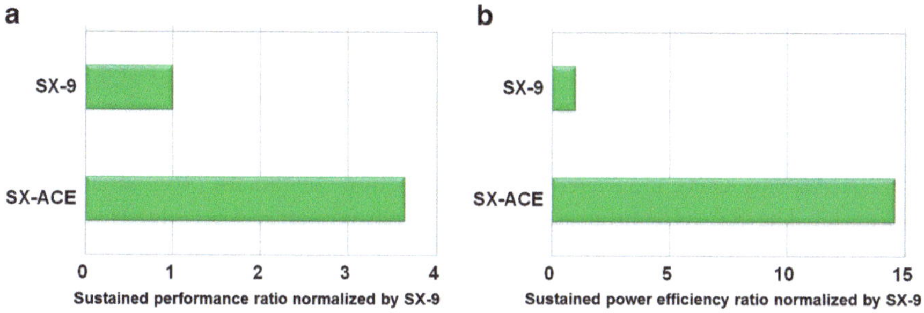

Fig. 11 Performance evaluation using one processor for the Himeno benchmark. (**a**) Sustained performance ratio, (**b**) sustained power efficiency ratio

Figure 10 shows the experimental results of sustained memory bandwidth evaluated by the STREAM TRIAD benchmark with the varying number of cores from one to four. The SX-ACE processor can provide approximately 220 GB/s sustained memory bandwidth in all cases. It is demonstrated that the SX-ACE processor is capable of fully utilizing its excellent bandwidth in contrast to current multi-core processors [9]. While the SX-ACE is designed on the basis of the multi-core architecture, even a single core of its processor can utilize the whole memory bandwidth.

We also evaluate the sustained performance of SX-ACE by using the Himeno benchmark program as shown in Fig. 11. In this evaluation, sustained performance and sustained power efficiency are compared between single processors of SX-9 (single core) and SX-ACE (four cores). Here the SX-9 processor having a single core provides a 102.4 GF theoretical peak performance with a 256 GB/s memory bandwidth. As shown in Fig. 11a, while the SX-9 processor has a same memory bandwidth as the SX-ACE processor (256 GB/s), the SX-ACE processor provides approximately 3.6× higher sustained performance in the memory-intensive Himeno benchmark program. The efficient memory subsystem of SX-ACE especially with

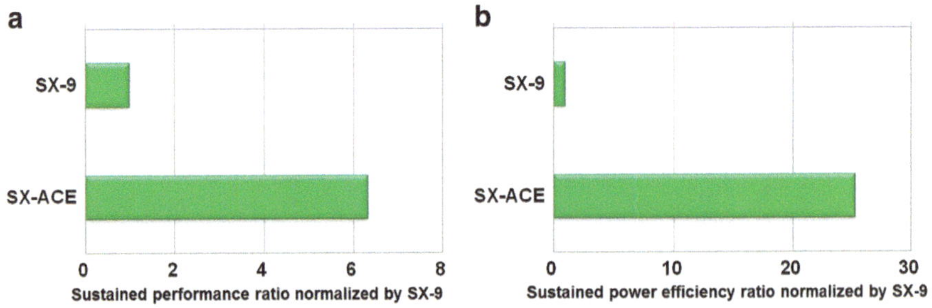

Fig. 12 Performance evaluation using one processor for the Legendre transformation kernel. (**a**) Sustained performance ratio, (**b**) sustained power efficiency ratio

the enhanced ADB capacity and the newly implemented MSHR function can effectively handle complex memory access patterns appearing in this benchmark program. Figure 11b shows the sustained power efficiency ratio which indicates a sustained performance per power consumption. The obtained power efficiency of SX-ACE is approximately 14.6× higher than the SX-9.

The sustained performance of indirect memory accesses is evaluated for the SX-9 and SX-ACE by using the Legendre transformation benchmark program as shown in Fig. 12. The SX-ACE processor provides approximately 6.3× higher sustained performance than the SX-9, as shown in Fig. 12a. The efficient memory subsystem with ADB, MSHR, and the out-of-order memory access accelerates accesses to the indirect memory in spite of the same memory bandwidth per processor. As shown in Fig. 12b, a higher power efficiency of approximately 25.3× than the SX-9 is obtained.

Conclusions

In this paper, we presented the brand-new vector supercomputer SX-ACE with the focus on its processor architecture, system architecture, and implementations. The described vector processor architecture is inherited from the previous SX Series models with the major design concept of a high performance core well commensurate with its large memory bandwidth. This is a key to providing higher sustained performance in real scientific applications. The SX-ACE core provides the worlds top-class performance of 64 GF with the largest memory bandwidth per core of 64–256 GB/s. The SX-ACE processor consists of four cores with its aggregate performance of 256 GF and memory bandwidth of 256 GB/s, realizing B/F ratio of 1.0. An efficient memory subsystem is introduced for enhancing sustained memory bandwidth by reducing redundant memory transactions. Experimental results utilizing elemental memory-intensive benchmark programs show the effectiveness of

(continued)

the memory subsystem employed by SX-ACE in eliciting superior sustained performance.

References

1. Satoshi, N, Satoru, T., Norihito, N., Takayuki, W., Akihiro, S.: Hardware technology of the SX-9 (1)-main system. NEC Tech. J. **3**(4), 15–18 (2008)
2. Takahara, H.: NEC SX Series Vector Supercomputer. Encyclopedia of Parallel Computing, vol. 4, pp. 1268–1277. Springer, Berlin (2011)
3. Soga, T., Musa, A., Shimomura, Y., Itakura, K., Okabe, K., Egawa, R., Takizawa, H., Kobayashi, H.: Performance evaluation on NEC SX-9 using real science and engineering applications. In: Proceedings of the Conference on High Performance Computing Networking, Storage and Analysis, pp. 1–12 (2009)
4. Zeiser, T, Hager, G., Wellein, G.: The worlds fastest CPU and SMP node some performance results from the NEC SX-9. In: Proceedings of IEEE International Symposium on Parallel & Distributed Processing (IPDPS2009), pp. 1–8 (2009)
5. Musa, A., Sato, Y., Soga, T., Egawa, R., Takizawa, H., Okabe, K., Kobayashi, H.: Effect of MSHR and prefetch mechanisms on an on-chip cache of the vector architecture. In: Proceedings of International Symposium on Parallel and Distributed Processing with Applications, pp. 335–342 (2008)
6. McCalpin, J.: Memory bandwidth and machine balance in current high performance computers. In: IEEE Computer Society Technology committee on Computer Architecture (TCCA) Newsletter, pp. 19–25 (1995)
7. The Himeno benchmark. http://accc.riken.jp/2444.htm
8. Sato, Y., Inoguchi, Y., Luk, W., Nakamura, T.: Evaluating reconfigurable dataflow computing using the Himeno benchmark. In: Proceedings of International Conference on Reconfigurable Computing and FPGAs (ReConFig), pp. 1–7 (2012)
9. Kerbyson, D.J., Barker, K.J., Vishnu, A., Hoisie, A.: Comparing the performance of Blue Gene/Q with Leading Cray XE6 and InfiniBand systems. In: Proceedings of 2012 IEEE 18th International Conference on Parallel and Distributed System, pp. 556–563 (2012)

SX-ACE, the Brand-New Vector Supercomputer for Higher Sustained Performance II

Noritaka Hoshi and Shintaro Momose

Abstract This paper describes an introduction of NEC's brand-new vector supercomputer, SX-ACE. The SX-ACE, which inherits and improves the NEC's SX vector architecture, is developed to provide superior sustained performance, especially for memory-intensive applications. For this purpose, the SX-ACE processor achieves the world's top-class single core performance of 64 GFLOPS and the world's leading memory bandwidth of 64 GB/s per core. Moreover, the SX-ACE system is designed to reduce power consumption to one-tenth with just one-fifth the floor space as compared to the previous SX-9 model by maintaining the same system performance. In this paper, we elaborate on the design concept, architectural overview, and implementation of the SX-ACE. We also discuss the sustained performance of the SX-ACE for several benchmark codes. The evaluation results demonstrate significantly improved sustained performance and power efficiency of the SX-ACE system in comparison with the SX-9.

1 Introduction

NEC made an announcement about the global launch of a brand-new vector supercomputer SX-ACE [1] in November 2013. It is the latest model of NEC's SX Series of vector supercomputers, featuring the top-level processor core performance and memory bandwidth. NEC has continuously been pursuing high sustained performance with the SX Series for the last 30 years. Since the launch of its pioneering systems SX-1 and SX-2 in 1983, the SX Series has been utilized worldwide in diversified areas, including academic research and industries. The SX-ACE, similar to all the other models of the SX Series, is a supercomputer aimed at providing both high sustained performance and ease of use.

In this paper, we describe its design concept, architectural overview, and implementation of the SX-ACE. Also, we demonstrate the results of performance evaluation for the SX-ACE.

N. Hoshi (✉) • S. Momose
NEC Corporation, Fuchu-city, Tokyo, Japan
e-mail: hoshi@ah.jp.nec.com; s-momose@ak.jp.nec.com

M.M. Resch et al. (eds.), *Sustained Simulation Performance 2014*,
DOI 10.1007/978-3-319-10626-7_6

2 Concept of Design

We have designed the SX-ACE with two concepts—one is the big core concept and the other is the reduction of power and space.

2.1 Big Core Concept

The SX-ACE processor is designed to satisfy both high peak performance and high memory bandwidth. In this paper, we call this concept 'Big core.' The following is the reason of this concept.

Firstly, we consider the trend of HPC systems. Fig. 1 shows the trend of top 10 systems on the TOP500 list [2]. While the LINPACK performance is increasing due to the growing numbers of cores and nodes per system, the single core performance and the processor frequency remain almost constant. It implies that growing LINPACK performance has been realized simply by the increase in system size, but it also causes increasing difficulty of programming. Users must often make tremendous efforts in gaining higher sustained performance with massive parallelism; however, achieving a parallelization ratio of close to 100 % is usually quite challenging for practical application programs. Therefore, we believe that the big core concept with smaller number of cores is an expected solution to enhanced usability without significant challenges resulting from parallel programming.

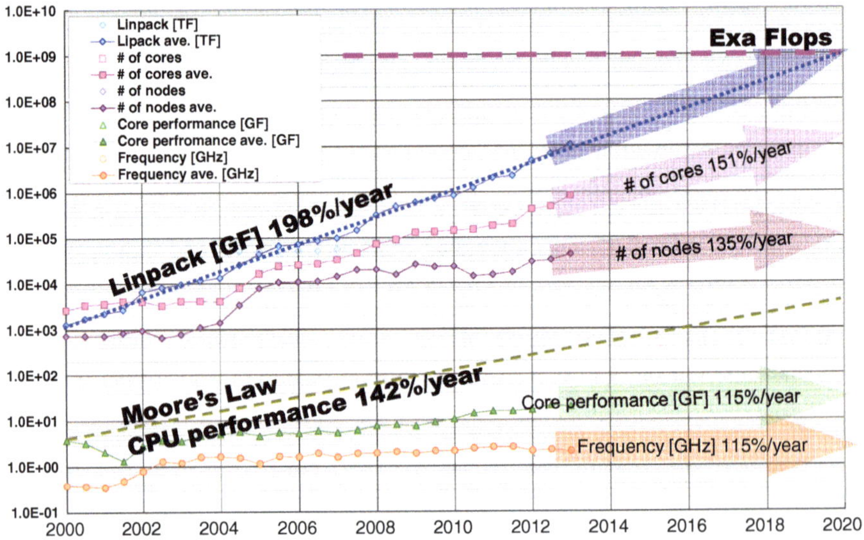

Fig. 1 Trends of top 10 systems in TOP500 list

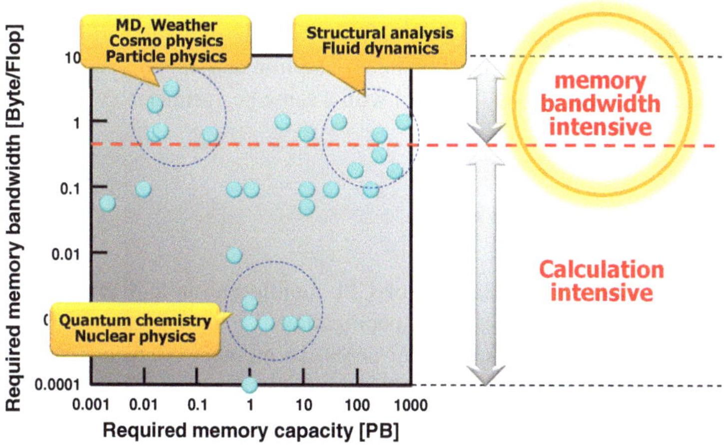

Fig. 2 Target applications of the SX-ACE

Secondly, we studied the characteristics of applications. According to a Japanese government (MEXT) working group's report for a wide variety of strategic applications, as shown in Fig. 2, diversified characteristics are observed [3]. In this figure, the x-axis indicates a required memory capacity and the y-axis a required memory bandwidth to run each application. The Bytes per Flop requirement (y-axis in this figure) of each application differs greatly. This report suggests that any single architecture can not cover all application areas. In particular, memory-intensive applications (e.g., molecular dynamics, cosmological physics, particle physics, structural analysis, fluid dynamics) need a larger memory bandwidth. In order to satisfy the desired memory capabilities for these types of applications, the SX-ACE processor is designed to realize a larger sustained memory bandwidth with its high Byte-per-Flop ratio.

On the basis of the big core design concept, SX-ACE provides the world top-class single core performance of 64 GFLOPS, as well as the world largest memory bandwidth of 64 GB/s per core.

2.2 Reduction of Power and Space

The single node of the SX-9 [4–7], which is a previous model of the SX Series and released in 2008, has 16 CPUs with each CPU having an LSI connected to a large shared memory system. In order to realize the large shared memory system, the single node of the SX-9 is equipped with 32 memory network switch LSIs and 512 memory controller LSIs, resulting in 560 LSIs in total. Such a large number of LSIs require high power consumption and a large installation space. Environment-friendly design is important also for a supercomputer and so we have improved power consumption and installation space. Using NEC's leading edge technologies for LSI design, high-density packaging, and high-efficiency cooling, SX-ACE realizes one-tenth the power consumption and one-fifth the floor space as compared to the previous SX-9 model with a same performance.

3 Architecture Overview

Figure 3 shows the block diagram of CPU configuration and Fig. 4 shows CPU floorplan. Table 1 summarizes CPU specifications.

The SX-ACE CPU consists of four cores, 16 memory controllers, one crossbar unit and an RCU (Remote access Control Unit). Four cores in the CPU share a 256 GB memory system in the SMP configuration. The crossbar unit, which connects four cores and 16 memory controllers, is located in the center of the CPU LSI. Each core can access 16 memory channels equidistantly. The average memory bandwidth is 64 GB/s per core; however, when only a single core is used, it can use the whole memory bandwidth of 256 GB/s.

Each core consists of SPU (Scalar Processing Unit), VPU (Vector Processing Unit), and ADB (Assignable Data Buffer), which is basically same as the SX-9 core configuration. ADB is a 1 MB data buffer located between the vector register and the memory system, designed to improve sustained memory bandwidth and shorten latency. The vector data with higher locality can be cached on ADB by specifying a directive in a source code. On the other hand, non-cached data can be controlled so as to bypass ADB.

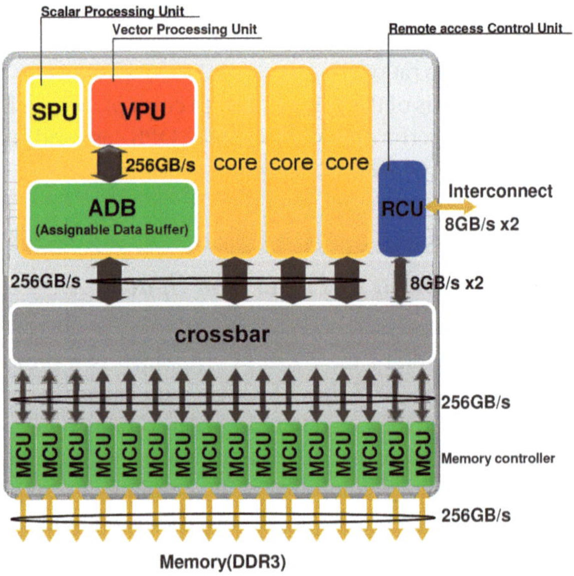

Fig. 3 CPU block diagram of SX-ACE

Fig. 4 CPU floorplan of SX-ACE

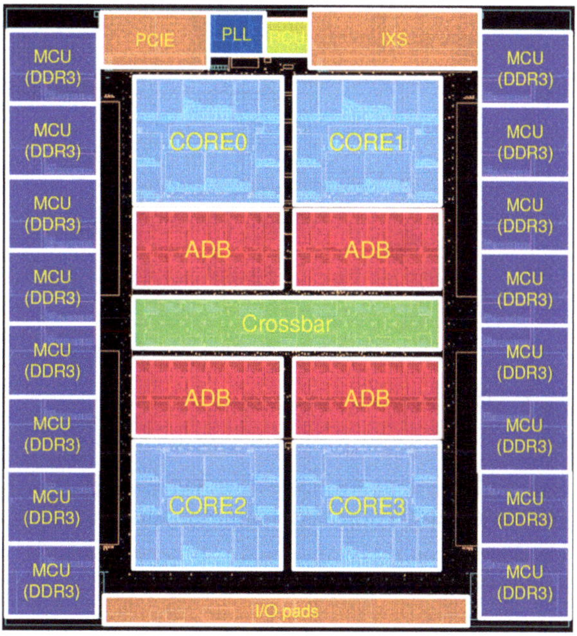

Table 1 CPU specifications of SX-ACE

Core	
Core performance	64 GFLOPS
ADB size/core	1 MB
ADB bandwidth	256 GB/s
Memory bandwidth/core	64–256 GB/s
Memory byte/flop	1.0–4.0
CPU	
# Of cores /CPU	4
CPU Performance	256 GFLOPS
Memory Bandwidth / CPU	256 GB/s (is provided by 16 DDR3 interfaces)
Byte/Flop	1.0
IXS bandwidth/CPU	8 GB/s × 2
IO I/F	PCIE(Gen2) × 8lane × 2port
Process rule	28 nm
Clock speed	1 GHz
Die size	23.05 × 24.75 mm
# Of transistors	2BTr.

The SX-ACE can be configured as a multi-node system by using a high-speed interconnect network named IXS. Each CPU is connected to IXS through RCU. RCU has 2 interface ports for a data transfer to/from IXS. Each port provides a 4 GB/s transfer rate per direction. Figure 5 illustrates an example of multi-node

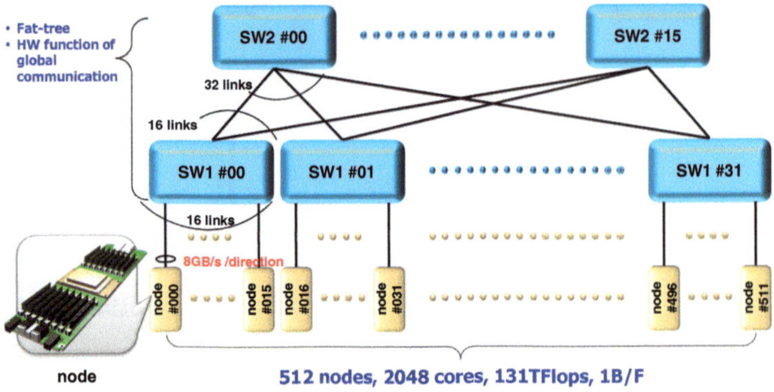

Fig. 5 512 nodes system of SX-ACE

system. In this figure, there are 512 nodes with the whole peak performance of 131TFLOPS. Global communication functions, global barrier mechanism, and communication registers which can be accessed from each node, and so on, are implemented into the hardware.

4 Implementation

Figure 6 is a picture of a single node card of SX-ACE. A CPU chip, 16 DDR3 memory modules, and network interface modules are mounted on this small 11 cm × 37 cm node card. The CPU chip is designed as an all-in-one processor, which includes four high-performance cores, 16 memory controllers, network controllers, and I/O interfaces. The all-in-one processor reduces the number of LSIs to one-hundredth compared with the SX-9, while high performance is maintained.

Figure 7 is a picture of a 2-node module, which includes 2 node cards and a water cooling module. It is a minimum unit for maintenance.

As a result, the configuration based on SX-ACE requires less power consumption and space than the SX-9 configuration for a same system performance. Figure 8 shows a comparison of size and power between the SX-9 and SX-ACE with the same peak performance, 131TFLOPS. The SX-9 configuration requires 80 racks and a 24 m × 12 m installation space. On the other hand, the SX-ACE configuration requires only 8 racks and a 7 m × 8 m installation space. In terms of power consumption, the SX-ACE configuration requires only 0.24 MW rated power consumption, which is one-tenth compared to SX-9.

Table 2 shows rack specifications. 64 nodes (or 64CPUs) are implemented in one rack with its performance of over 16TFLOPS and memory bandwidth of over 16 TB/s.

Fig. 6 Single node card of SX-ACE

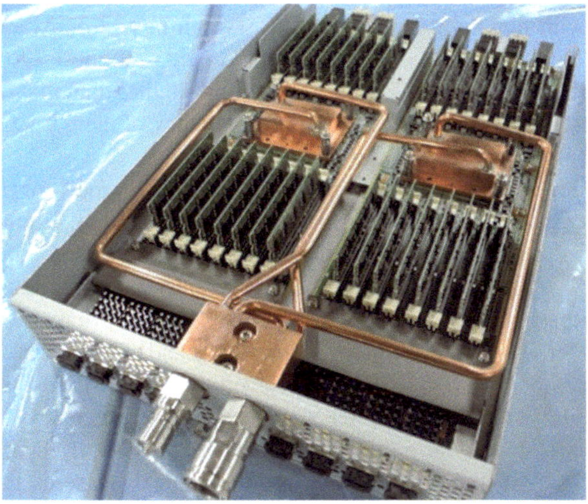

Fig. 7 2-Node module of SX-ACE

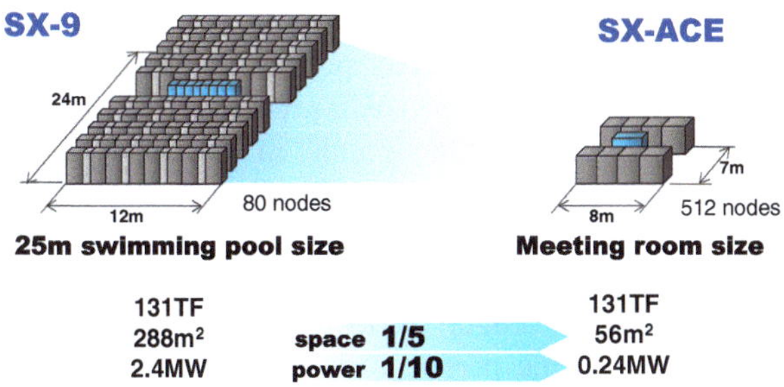

Fig. 8 Downsizing and power saving

Table 2 Rack specifications

# Of nodes	64
Peak performance	16.3TFLOPS
Total memory bandwidth	16.3 TB/s
Rated power consumption	30 KW
Size W × D × H (m)	0.75 × 1.2 × 2.0

5 Performance Evaluation

In this section, we demonstrate the results of performance evaluation of the SX-ACE for several applications. We evaluate sustained memory bandwidth, short vector performance, and indirect memory access performance.

At first, we evaluate sustained memory bandwidth and power efficiency for the STREAM TRIAD benchmark and the Himeno benchmark. Figure 9 shows the evaluation result for the STREAM TRIAD benchmark. Here the SX-ACE CPU is compared with the Xeon Ivybridge CPU, which has 12 cores and a 51.2 GB/s memory bandwidth. The SX-ACE processor provides about a 220 GB/s memory bandwidth for one to four cores, which is 26 times higher than that of a single core of the Xeon processor. The sustained power efficiency of the SX-ACE processor is 1.9 times higher than that of the Xeon processor.

Figure 10 shows the comparison between the SX-ACE processor (four cores) and the SX-9 processor (one core) using Himeno benchmark [8]. Himeno benchmark is designed to evaluate the performance in solving the Poisson equation with the Jacobi iterative method as a typical memory-intensive benchmark program. The SX-ACE processor (4 cores) and the SX-9 processor (1 core) have the same memory bandwidth of 256 GB/s. In this figure, performance and efficiency are normalized to the SX-9 processor. The SX-ACE processor provides a 3.6 times higher sustained performance and a 15 times higher sustained power efficiency compared with the SX-9 processor. This is partly due to the enlarged ADB capacity (1 MB per core) of the SX-ACE processor, which is four times as large as the ADB of the SX-9 processor. Furthermore, ADB has a mechanism for avoiding redundant loads (MSHR: Miss Status Holding Registers) and MCU (Memory Controller Unit) has a mechanism for avoiding redundant stores. These features contribute to achieving higher sustained memory bandwidth than the SX-9.

Secondly, we evaluate the short vector performance, which is improved from the SX-9 processor in designing the SX-ACE processor by the enhancement of the bypass chaining path between arithmetic units in VPU with the reduced latency of vector operations. In addition, the vector instruction issuing function in SPU is also enhanced, accelerating the short vector performance as shown in Fig. 11. Here we selected six in-house benchmark codes for evaluating the short vector performance with the vector length set to 32. Figure 11a shows the measured performance normalized to the performance on the SX-9 for each kernel. The sustained performance of the SX-ACE processor is 1.3 times higher than that of the SX-9 on the average, even though the theoretical single-core performance of

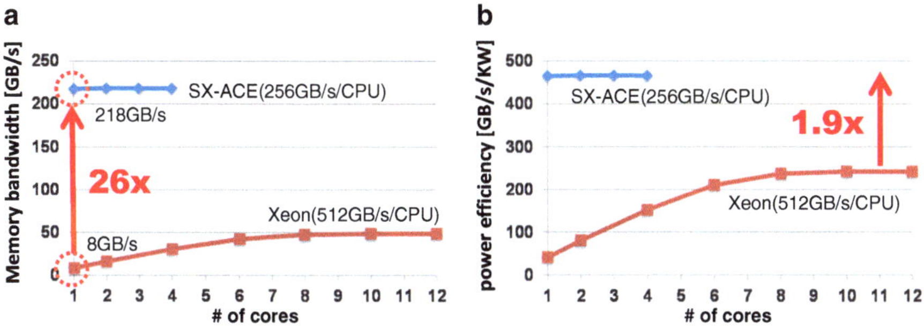

Fig. 9 Sustained memory bandwidth majored by STREAM TRIAD. (**a**) Sustained memory bandwidth, (**b**) sustained power efficiency

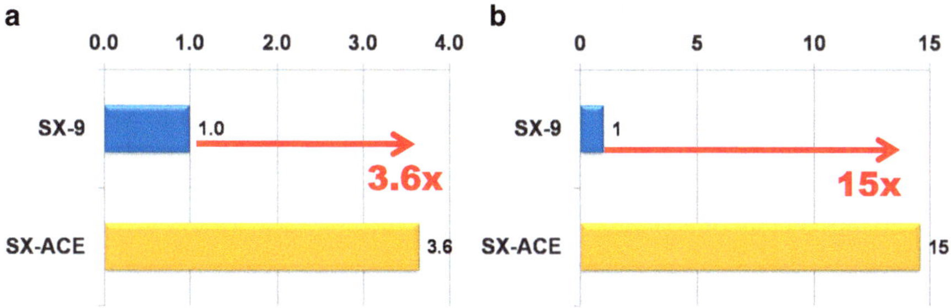

Fig. 10 Sustained performance majored by Himeno benchmark. (**a**) Sustained performance normalized by SX-9, (**b**) sustained power efficiency normalized by SX-9

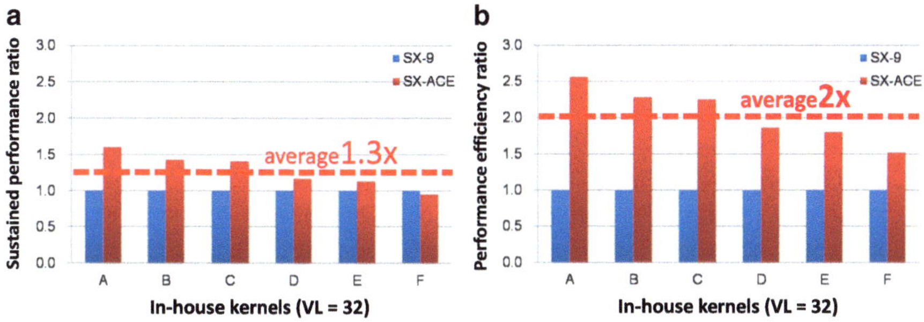

Fig. 11 Short vector performance. (**a**) Sustained performance, (**b**) Sustained performance efficiency

the SX-9 processor is 102.4 GFLOPS, which is higher than that of the SX-ACE processor (64 GFLOPS). Figure 11b indicates the sustained performance efficiency ratio of each kernel normalized to the SX-9. As shown in this figure, the obtained sustained performance efficiency ratio of the SX-ACE is 2 times higher than the SX-9 on the average.

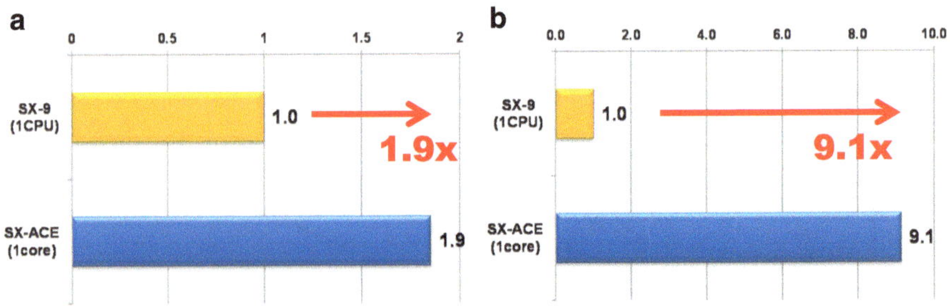

Fig. 12 Indirect memory access performance. (**a**) Sustained performance normalized by SX-9, (**b**) sustained power efficiency normalized by SX-9

Indirect memory access is also improved in comparison with the SX-9 processor. By eliminating network controller LSIs, integrating MCU into a single chip, and enlarging the ADB capacity, the memory access latency of SX-ACE is shortened in comparison with the SX-9. The instruction reordering function in SPU is also enhanced, resulting in the improved indirect memory access. Figure 12 shows the evaluated result for the Legendre transformation program, which often appears in scientific computing such as numerical weather forecast. Here the single core performance of the SX-9 processor is compared with the single core of the SX-ACE processor. These cores have the same memory bandwidth of 256 GB/s, while the theoretical single-core performance of the SX-9 is 102.4 GFLOPS, which is higher than the SX-ACE. Figure 12 shows the sustained performance (a) and the sustained power efficiency (b) normalized to the SX-9. The sustained performance of the SX-ACE is 1.9 times as higher as the SX-9. Furthermore, in spite of the same memory bandwidth, the power consumption of SX-ACE is significantly reduced against the SX-9.

Conclusion

In this paper, we gave an introduction of NEC's brand-new vector supercomputer, which was released in November 2013. SX-ACE, which inherits and improves the NEC's SX Series vector architecture, is developed to provide superior sustained performance especially for memory-intensive applications. For this purpose, the SX-ACE system is based on a big core concept with an environment-friendly design. In line with such a policy, the SX-ACE processor has achieved the world's top-class single core performance of 64 GFLOPS and a memory bandwidth of 64 GB/s/core. Moreover, the SX-ACE system reduces power consumption to one-tenth and requires just one-fifth the floor space of the previous SX-9 model with the same performance. This paper also discussed the sustained performance of SX-ACE for several bench-

(continued)

mark codes, demonstrating significantly improved sustained performance and power efficiency of the SX-ACE system in comparison with the SX-9.

References

1. Momose, S., Hagiwara, T., Isobe, Y., Takahara, H.: The Brand-New Vector Supercomputer, SX-ACE. In: International Supercomputing Conference (ISC 2014), pp. 199–214 (2014)
2. Top500 supercomputing sites. http://www.top500.org/ (2013)
3. Working group of the Ministry of Education, Culture, Sports, Science and Technology of Japan, White Paper for Strategic Direction/Development of HPC in Japan (2012)
4. Satoshi, N., Satoru, T., Norihito, N., Takayuki, W., Akihiro, S.: Hardware Technology of the SX-9 (1) Main System. NEC Tech. J. **3**(4), 15–18 (2008)
5. Takahara, H.: NEC SX Series Vector Supercomputer. In: Encyclopedia of Parallel Computing, vol. 4, pp. 1268–1277. Springer, Berlin (2011)
6. Soga, T., Musa, A., Shimomura, Y., Itakura, K., Okabe, K., Egawa, R., Takizawa, H., Kobayashi, H.: Performance Evaluation on NEC SX-9 using Real Science and Engineering Applications. In: Proceedings of the Conference on High Performance Computing Networking, Storage and Analysis, pp. 1–12 (2009)
7. Zeiser, T., Hager, G., Wellein, G.: The world's fastest CPU and SMP node: Some performance results from the NEC SX-9. In: Proceedings of IEEE International Symposium on Parallel & Distributed Processing (IPDPS 2009), pp. 1–8 (2009)
8. The Himeno benchmark. http://accc.riken.jp/2444.htm (2013)

Feasibility Study of a Future HPC System for Memory Intensive Applications: Conceptual Design of Storage System

Ken'ichi Itakura, Akihiro Yamashita, Koji Satake, Hitoshi Uehara, Atsuya Uno, and Mitsuo Yokokawa

Abstract We started a feasibility study of high-end computing systems as a 2-year national project in 2012 for exa-scale computing era. In order to realize the exa-scale system, it is extremely important to design a large scale storage system with high bandwidth.

We made a conceptual design of a mass storage system with high-speed I/O technology by studying future I/O requirements from simulations for disaster prevention and mitigation research. We concluded that a distributed and parallel file system with multiple meta-data servers is required, because a meta-data server (MDS) using in the Lustre file system is apparently insufficient to handle a large scale file system. In this paper, the storage system architecture and its performance are presented.

1 Introduction

We started a feasibility study of high-end computing systems as a 2-year national project in 2012 for exa-scale computing era [1]. In this feasibility study, we studied a conceptual design of an exa-scale supercomputer system which is required around 2018 to contribute making safe society and bringing forth new strong

K. Itakura (✉) • H. Uehara
JAMSTEC, Yokohama, Japan
e-mail: itakura@jamstec.go.jp; uehara@jamstec.go.jp

A. Yamashita • K. Satake
NEC, Tokyo, Japan
e-mail: a-yamashita@aj.jp.nec.com; k-satake@vf.jp.nec.com

A. Uno
RIKEN, Kobe, Japan
e-mail: uno@riken.jp

M. Yokokawa
RIKEN/Kobe University, Kobe, Japan
e-mail: yokokawa@port.kobe-u.ac.jp

M.M. Resch et al. (eds.), *Sustained Simulation Performance 2014*,
DOI 10.1007/978-3-319-10626-7_7

industry technology for national competitiveness. Principal target applications in our feasibility study are the fields of disaster prevention and mitigation.

In the beginning of the study, we clarified social and scientific issues in the fields, and we made an application development road map. Based on the road map and trends of device technology, the state of the HPC system architecture was surveyed and the conceptual design was performed. The most important issue was power consumption. We focused on not only the peak computational performance per power consumption but also a sustained performance of applications per power consumption. The system was designed so that the sustained performance was made as high as possible.

Several teams are organized to examine parts of the HPC system in the project. We made a conceptual design of a mass storage system with high-speed I/O technology. In this paper, we describe the storage system architecture and performance.

2 Objectives

2.1 Design Cycle

The design cycle is important to make conceptual design of a large scale storage system with high I/O performance for big data such as observation data and simulation data appeared in disaster prevention and mitigation simulations.

We determined a storage system is a hierarchical structure as a base architecture to keep system scalability in consideration of cutting-edge technology.

First of all, we analyze I/O behaviors of applications in order to decide requirements for I/O performance. The design parameters are drawn by the information of I/O data amount, the number of I/O, frequency, and ratio of I/O time to the total execution time. Once the conceptual design is made, I/O performance of applications is estimated. Then we consider the data structure of applications and the algorithm used in applications should be changed to obtain higher performance. Then we go back to the first process. This is a design cycle.

2.2 Requirements from Applications

In this feasibility study, the requirements to the computer architecture were considered with scientific accuracy and computational complexity of the many target applications.

First, the program organization and performance requirement of each target application were analyzed. Next, operation counts, required main memory capacity, required memory throughput (Byte/Flops), data I/O, execution time were extracted.

Especially, the system requires high-speed I/O technology and mass storage system for processing observation and simulation huge data of the disaster prevention and mitigation study efficiently.

Here, the number of compute nodes, the capacity of file system and peak I/O bandwidth were premised 10–200 K, 250–1,000 PB and 10–30 TB/s, respectively, to examine the storage system. The capacity and performance should be changed according to the scale of a target system. The target system is also considered to be equipped with a vector processing mechanism with a high memory bandwidth, high operation performance, high node performance, a low network latency, and mass storage.

As a result of performing a fundamental quality assessment, it turned out that high effective performance system is obtained. Each component of the system is designed aiming at this architecture.

In order to realize such a large scale and high bandwidth storage system, a distributed and parallel type file system organization by many storage devices are required. However, by the method of managing files by a meta-data server (MDS) like a Lustre file system [2], MDS serves as a bottleneck. A file system with multiple meta-data servers which distributed over the file system is required.

In order to improve reliability, availability, and serviceability (RAS) of storage system, organizations adaptable to intermittent faults such as duplicated access paths and a data restoration function are required. Required storage capacity should be determined from main memory capacity and bandwidth between memory and the storage.

Although the storage capacity should be determined by the requirements from applications, the capacity is limited by the restrictions of a system scale and development cost. It is not realistic to save all calculation results in a time evolution simulation, and therefore, minimum storage capacity based on the amount of minimum demands of applications should be kept. In addition, file I/O capacity, I/O time and I/O frequency were examined from the demand of the comprehensive disaster prevention simulation which is one of the target applications.

3 Design Concept and Result

High bandwidth can be realized by increasing and parallelizing the amount of resources. However, each part of the system requires the minimum bandwidth and capacity.

From the design results of a compute node with an interconnect network, a base model of the system consists of 25,000 compute nodes and 8 compute nodes share an I/O node. To fit a storage system to the compute system, a two-level storage system consisting of local storage of solid state disks (SSDs) and global storage system of conventional hard disk drives (HDDs) is needed.

The local SSD storage system has good response performance. The global HDD storage is connected to highly parallel and throughput storage area network (SAN).

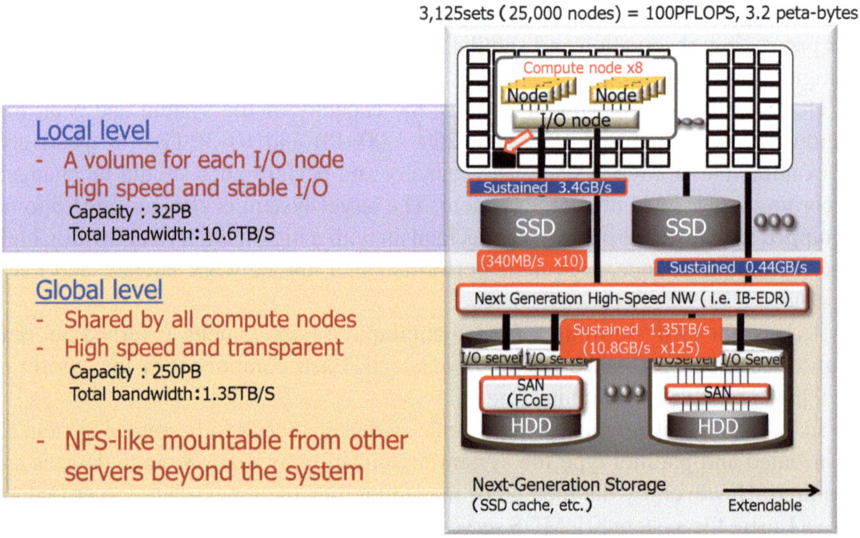

Fig. 1 A design of storage system which is a distributed and hierarchical system. The total capacity is 282 PB

The first level local storage provides stable I/O performance and is shared by 8 compute nodes. Use cases of the local storage system are temporary use, file staging, check pointing, and cache storage of the second level storage. The second level global storage is provided high bandwidth for all calculation nodes and shared by all nodes. Use case of the global storage is to store the permanent files for the huge I/O data from/to a large number of nodes.

After deciding the base model, the storage system became the distributed shared hierarchical type shown in Fig. 1.

The first level storage system is served as an I/O server. It has 32 PB capacity in total and 10.6 TB/s whole throughput performance. The second level global storage has 250 PB capacity. The storage area network (SAN) between the I/O nodes and I/O servers consists of infiniband (EDR) and the throughput performance is 1.35 TB/s. The high I/O throughput from all the compute nodes is realized. Moreover, mounting a file system on the other servers which are located in the SAN can be enable.

The global distributed parallel file system should be equipped scalability performance, fault tolerance of the file system with MTBF, consistency of meta-data and data body in the file access transaction, and end-to-end data consistency in read/write operations. The technologies to be developed are shown in Fig. 2.

This system should have backup redundancy of an access path and data restoration function on a course in order to avoid the stop by a single point of failure. The duplicate management to multiple MDSs about meta-data with software RAID technology can attain the high performance for failures.

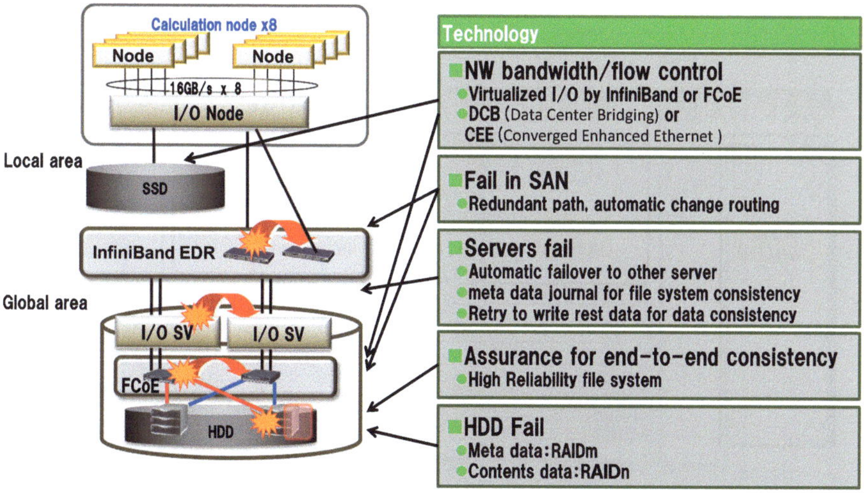

Fig. 2 RAS for the storage system. It supports the single-point-of-failure exclusion, automatic recovering and keeping consistency of the file system

The duplicated meta-data organization can maintain high scalability, when the hot spot of accesses to the meta-data occurs and is obviously effective in management of the hundreds of thousands of files which target applications create. The meta-data access operations should be extended so that the files placed in parallel can be treated systematically. This reduces the number of transactions and client processes. By these capabilities, a large-scale file system can be employed with high reliance.

4 Storage System Performance

We evaluated I/O time of the target applications assuming the applications are executed on the proposed system as well as with the designed storage system in cooperation with the application group.

A coupling simulation with several simulation codes for disaster mitigation needs high I/O bandwidth to store the output data into global storage. Figure 3 shows the execution flow and the analyzing result. The total I/O data capacity is 540.4 TB and the time required for I/O is 4.5 % of the total execution time of three hours.

A flow of the data assimilation simulation is shown in Fig. 4. This application consists of the forward time evolution parts and the backward parts. Each part shares the temporary data file which has to store to local file system. And this simulation initially needs the weather and climate observational data which are so large amount. The time required for the I/O is 2.3 % of the total execution time 4 h. Both results have achieved the target I/O time which is less than 10 % of all the execution time.

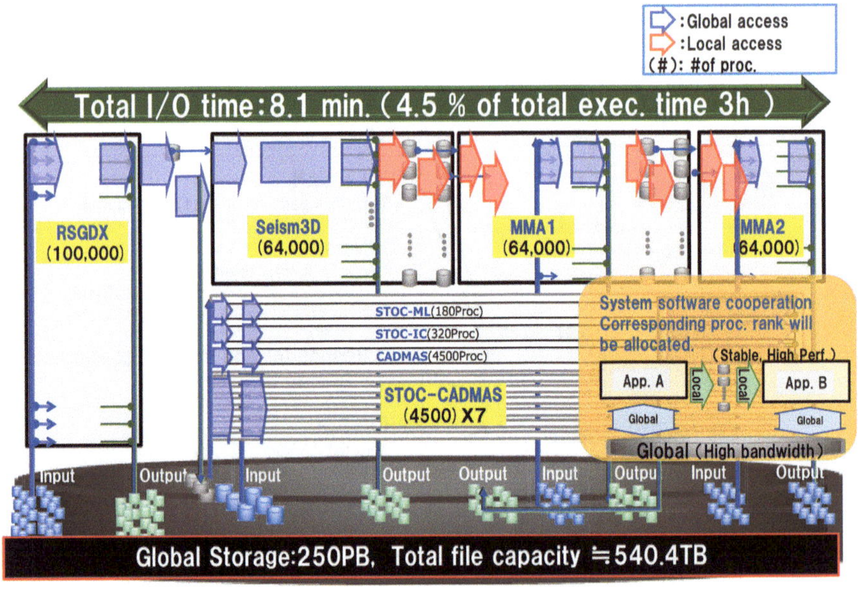

Fig. 3 Coupling simulation for disaster mitigation: high I/O bandwidth to store the output data into global storage

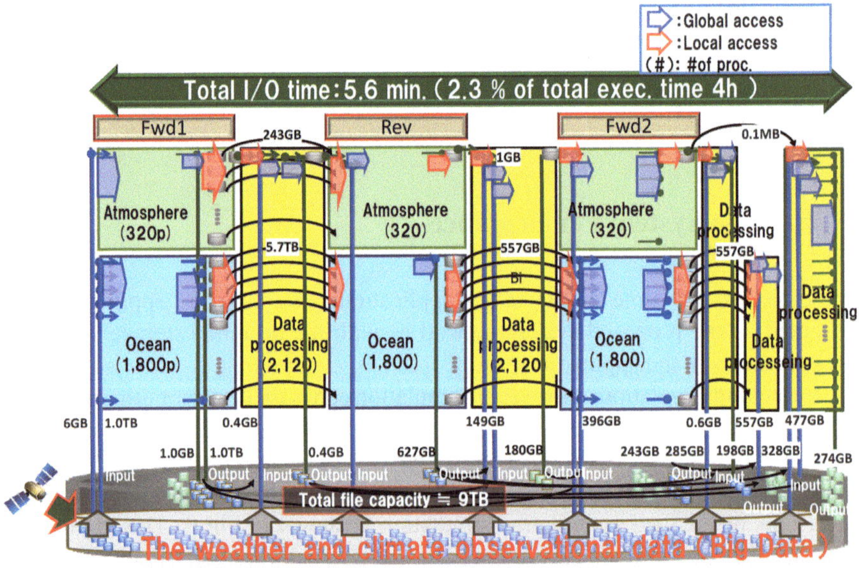

Fig. 4 Data assimilation: high I/O performance to store temporally shared data between applications in the local level storage

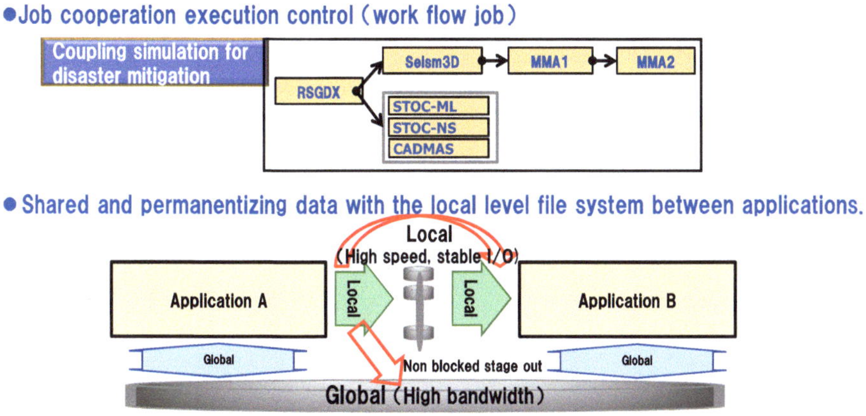

Fig. 5 System software cooperation

Distributed meta-data file system architecture has several advantages; access concentration to the same directory can be avoided. A failure of the meta-data service can be avoided. A response of file utility commands like "ls" or "du" can be reduced. In order to use the hierarchical storage system effectively, cooperation between the system software and the storage system should be well-organized.

A consecutive execution of a series of applications on the same compute node group is performed efficiently. When intermediate files of a series of applications are located on the local files system, the execution time can be shortened (Fig. 5).

5 Summary

The conceptual design of the mass storage system and high-speed I/O technology was performed by analyzing the future large-scale I/O data requirements with respect to the disaster prevention and mitigation simulation.

A coupling simulation and a data assimilation simulation are subsided within allowed time on two-level storage system architecture. It is desirable to evaluate the file system software on a prototype system at an early stage. It is required to re-design the storage system according to future technologies, I/O configuration in nodes and the budget.

Acknowledgements The authors wish to thank Prof. Hiroaki Kobayashi of Tohoku University for leading this project. This project is supported by Ministry of Education, Culture, Sports, Science and Technology of Japan.

References

1. Kobayashi, H.: Feasibility study of a future HPC system for memory-intensive applications: final report. In: Sustained Simulation Performance 2014. Springer, Berlin (2014)
2. "Lustre": http://wiki.lustre.org (2014)

Part II
Exploitation of Existing HPC Systems: Potentiality, Performance and Productivity

Designing an HPC Refactoring Catalog Toward the Exa-scale Computing Era

Ryusuke Egawa, Kazuhiko Komatsu, and Hiroaki Kobayashi

Abstract Aiming at keeping performance portability of practical applications even in the exascale computing era, this paper presents a concept of HPC refactoring, and the design of the HPC refactoring catalog. The HPC refactoring catalog provides a guideline to keep performance portability across different supercomputing systems based on the quantitative analysis of practical simulations. In this paper, our recent research activities for designing the HPC refactoring catalog are introduced.

1 Introductions

Recently, high-performance computing (HPC) systems become requisite facilities to accelerate various kinds of research in sciences and engineering fields. Over the past few decades, in response to the demand from computational scientists, the performance of HPC systems have been significantly improved. The world fastest supercomputer, in June 2014, achieves 33.8 PFlop/s in the LINPACK benchmark [1]. Moreover, toward designing exascale computing systems, computer architects are continually pursuiting in the race to design faster and energy efficient supercomputers. According to these performance improvements, the systems are getting complicated and diversified.

In general, the lifetimes of HPC application codes tend to be long because the codes have been incrementally improved along with the evolutions of HPC systems [2]. While HPC programmers can obtain extremely high computational performance given by the improvement, it is getting harder to exploit the potentials of HPC systems due to their complexity and variety. Since HPC programmers are just concerned with getting the most performance from existing HPC systems, applications have been also evolved for a long time by being optimized step by step

R. Egawa (✉) • K. Komatsu
Cyberscience Center, Tohoku University/JST CREST, 6-3 Aramaki-aza-aoba, Aoba, Sendai 980-8578, Japan
e-mail: egawa@isc.tohoku.ac.jp; komatsu@isc.tohoku.ac.jp

H. Kobayashi
Cyberscience Center, Tohoku University, 6-3 Aramaki-aza-aoba, Aoba, Sendai 980-8578, Japan
e-mail: koba@isc.tohoku.ac.jp

M.M. Resch et al. (eds.), *Sustained Simulation Performance 2014*,
DOI 10.1007/978-3-319-10626-7_8

for available supercomputers. However, current drastic changes in the complexity and variety of the HPC systems force to HPC programmers to further spend an enormous time to develop and migrate their codes toward the future systems.

Under this situation, aiming at realizing smooth and practical code-migrations toward the future HPC systems, this paper presents a concept of HPC refactoring, and the design of an HPC refactoring catalog. The HPC refactoring catalog presents the guidelines for keeping performance portability across different HPC systems, by identifying system-specific performance optimizations in HPC applications.

The rest of this paper is organized as follows. Section 2 describes the performance portability issues in high-performance computing and the concepts of HPC refactoring. Our activities for designing the HPC refactoring catalog are introduced in Sect. 3. Section "Conclusions" concludes this paper.

2 Performance Portability of HPC Applications

To keep maintainability and readability of a code, refactoring techniques are commonly used in software engineering fields. "Refactoring" is a disciplined technique for restructuring an existing body of codes, altering its internal structure without changing its external behavior [3]. A refactoring catalog describes guidelines to clarify how a code should be refactored. Since "refactoring" has less importance on performance, it works for improving readability, portability, and reusability of the code.

However, in HPC fields, the highest priority has been always given to performance. Therefore, existing HPC applications have been already optimized for a particular HPC system in many cases. Such system-specific optimizations significantly deteriorate the performance on different systems, and this fact also makes the migration of their codes much difficult. Figure 1 shows the impacts

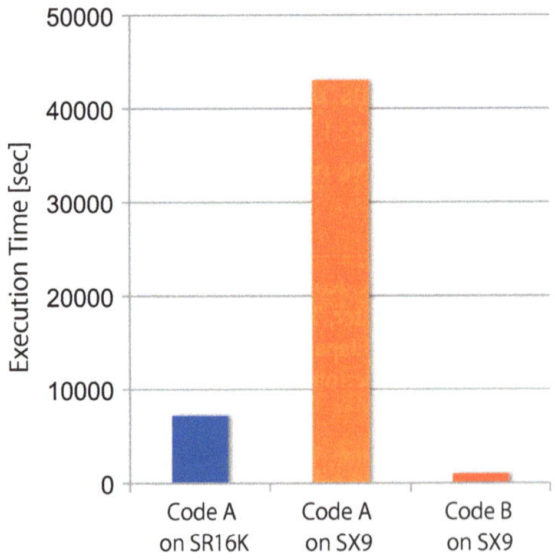

Fig. 1 A performance portability of practical application

of system-specific optimizations on performance. The vertical axis shows the execution time of the codes, and the horizontal axis indicates the HPC systems and codes. In this evaluation, two codes A and B with the same functions are used; "Code A" is optimized for SR16K and "Code B" is optimized for SX-9.

When running "Code A" on SX-9, the execution time becomes larger by a factor of six than those of SR16K. However, the performance of "Code B" on SX-9 achieves higher performance compared to "Code A" on SR16K. To generate "Code B," the system-specific optimizations for SR16K are identified and removed, then optimizations (loop-unrolling and remove redundant calculations) for SX-9 are carried out. From this example, we can easily confirm the importance of identifying the system-specific optimizations from the original code.

The aim of HPC refactoring is to keep a performance portability high, rather than readability and maintainability of the codes. Therefore, here, we define that HPC refactoring is the technique to keep a high performance portability by identifying the system-specific parts of the code. This identification enables to separate system-specific and non system-specific descriptions in the code. To realize HPC refactoring, the guidelines for carrying out this procedure is needed. The catalog should contain the cases which deteriorate performance portability by system-specific optimizations, and the detailed designs of the catalog are described in the following sections.

3 Designing an HPC Refactoring Catalog

3.1 Design Concepts

This section describes a design strategy and design activities of the HPC refactoring catalog. While, as we discussed in the introduction, heterogeneity and complexity of HPC systems are drastically increasing, the data processing mechanism itself does not drastically change. In addition, all systems tend to be headed in the same design direction. For example, recent scalar-based systems try to enhance vector processing mechanisms, accelerator architectures try to enhance their scalar performance with their innate SIMD processing mechanisms, and the current vector system also employs multicore architecture with on-chip memories [4]. Thus, we believe that elemental optimization techniques still have the capability to exploit the performance of the future system.

Based on this assumption, the overall flow for designing the HPC refactoring catalog is depict in Fig. 2. The flow can be divided into four steps.

Step 1 Optimize or migrate practical codes to a particular system, and gather optimized codes for a particular HPC system
Step 2 Accumulate and analyze the existing practical application codes
Step 3 Evaluate the codes and detect typical patterns
Step 4 List up the necessary conditions for keeping performance portability

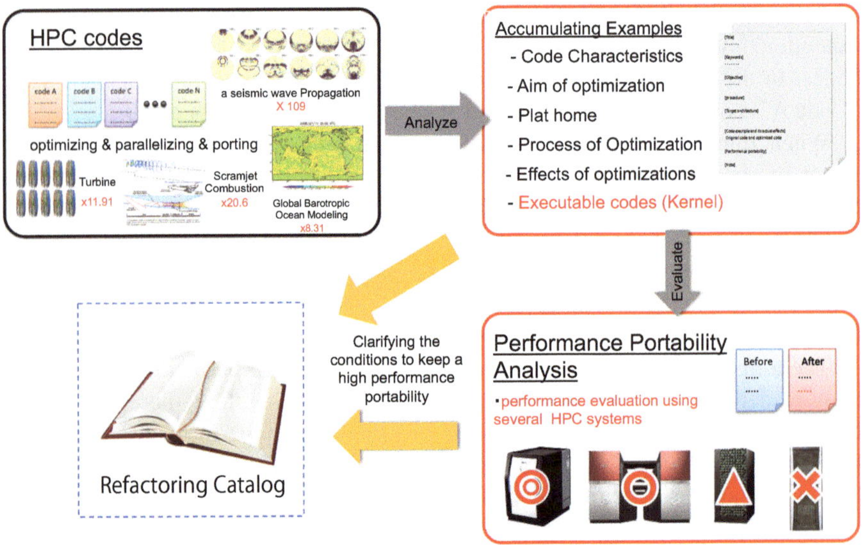

Fig. 2 Designing the HPC refactoring catalog

In Step 1, practical application codes are optimized and migrated to a particular system. Along with Step 1, practical application codes, which have been running on the NEC vector supercomputer SX series at Cyberscience center of Tohoku University, are analyzed in Step 2. In this step, the "lists of optimizations" are edited by classifying codes by optimization techniques applied to the codes. In this list, the following information of the optimization is accumulated: *objectives, procedures, keyword, target architecture, code examples* ***(executable kernels)***. The executable kernels include an original (not optimized) kernel, and optimized kernel are collected to enable multi-platform evaluations.

In Step 3, the performance of the executable kernel is evaluated on various HPC systems to clarify the effects of optimizations. Then performance portability of the codes are analyzed. In our work, performance portability is judged in the following approach. By evaluating the executable kernels, the performance improvement rate can be obtained by both execution times of original and optimized codes. Assume that the kernels are evaluated on four HPC systems; **A, B, C**, and **D**. In this case, as shown in Fig. 3, if performance improvements can be achieved in "all" the systems, we evaluate the performance portability of the optimization is high. On the other hand, if the performance improvement can be just observed in the limited number of systems, the performance portability of the optimization is judged as low. Based on this observation, the system-specific optimizations can be detected. Especially, if the performance improvements can be observed in one system, the optimization should be listed on the HPC refactoring catalog in Step 4.

To accumulate examples of optimizations, practical applications should be collected and analyzed. Since Cyberscience center has started joint research activ-

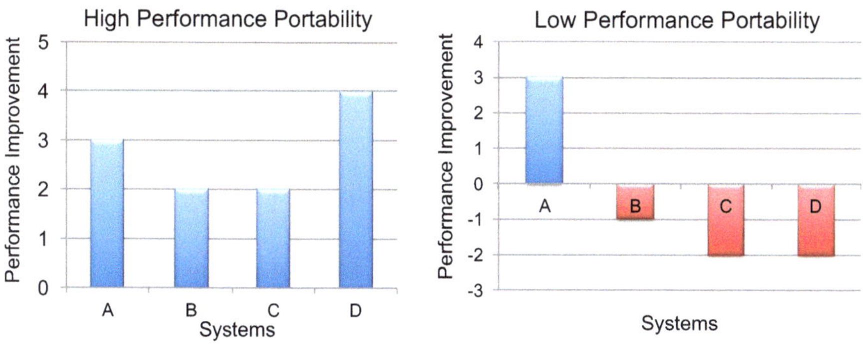

Fig. 3 Performance portability

Year	1997	1998	1999	2000	2001	2002	2003	2004	2005	2006	2007	2008	2009	2010	2011	2012
Number of optimized codes	2	8	8	9	10	7	18	20	8	16	10	15	8	8	13	6
Node performance improvement	1.9	46.7	4.5	2.5	1.6	2.2	6.7	2.9	1.3	2.9	33	9.4	381	47	16.2	19.7
Parallel performance improvement	11.1	18.4	31.7	8.6	4.9	2.8	18.6	4.5	5.3	8.1	1.9	5.1	3.6	48	17.2	15.3

Fig. 4 User support activities on Cyberscience center

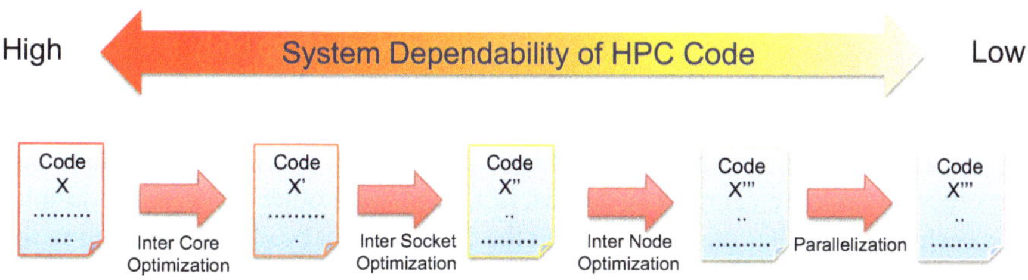

Fig. 5 General optimization flow

ities with computational scientist from 1997, we have started to analyze these applications, which have been running on the Center. Figure 4 shows the number of joint researches, and the results of code optimization activities in Cyberscience center. So far, almost 150 applications codes are optimized through the joint research activities.

Although there are many optimization techniques in several levels, i.e. core, socket, node, inter-node, as shown in Fig. 5, this study focuses on the inner-node level optimizations. Because, in general, an HPC application code is firstly optimized to improve its single-node performance, assuming a particular core, socket, and node architecture. Then the code is optimized for multi-node execution. In addition, comparing the inner-node level with the intra-node level, the optimizations applied to the inner-node level tend to be a highly system dependent. This inner-node level performance portability is becoming more important because the node architectures are diverging by having deeper memory hierarchies, and also increasing their variations and heterogeneity. Therefore, this work has been started with analyzing optimization techniques in the former level.

3.2 Current Status of the HPC Refactoring Catalog

In this study, as a first step for designing the HPC refactoring catalog, practical applications, which have been developed at Cyberscience center are analyzed. Through the analysis and classifications, 22 cases have been accumulated in the optimization lists. The performance evaluations are carried out using HPC systems listed in Fig. 6. Although there are many kinds of HPC systems, five representative systems are selected as evaluated platforms. These systems have their own system and architectural characteristics and their dedicated compilers, and then the performance evaluations of the executable kernels are carried out.

So far, we have detected four typical system-specific optimizations, which deteriorate performance portability; *(1) loop fusion, (2) loop unrolling, (3) cache blocking, (4) conversion of "structure of array" to "array of structure."* These four optimizations are useful for improving the performance on a particular system, however the performance are degraded in other systems. Figure 7 shows the pseudo code of Case (1). This code is optimized for SX-9, and in this code, triple nested loops are fused to a single loop. By this optimization, the loop body is enlarged and then the number of calculations which can be processed by a single instruction is increased. Therefore, this optimization is effective for the vector supercomputer SX-9, and a 12.6 $\times$ performance improvement can be achieved. However, performance on other systems are degraded as shown in Fig. 8. Since other four systems employ the cache-based scalar architecture, and the cache cannot be effectively used in this case. As same as this example, other three optimizations are effective for a particular system, but not for other systems. Thus, we can confirm that these optimizations have low-performance portability, and are system-specific optimizations.

System	Peak Gflops/s	Nodes	Sockets / node	Cores /PE	Memory BW GB/s	On-chip Memory	Network	B/F
NEC SX-9	1676.8	1	16	1	256	256 KB ADB	2x 128GB/s IXS	2.5
Nehalem EX	289.92	1	4	8	34.1	256 KB L2/core, 24MB shared L3	2x 4 GB/s IB	0.47
Fujitsu FX1	41.28	1	1	4	40	6MB shared L2	2x 4 GB/s IB	1.0
Fujitsu FX10	236	1	1	16	85	12 MB shared L2	2x 5 GB/s IB	0.36
Hitachi SR16K M1	980.48	1	4	8	128	256 KB L2/core, 32MB shared L3	2x 24 GB/s	0.52

Fig. 6 System specifications

Before

```
do k=2,nz-1
  do j=2,ny-1
    do i=2,nx-1
      ...
    end do
  end do
end do
```

After

```
do i=2*2*2,
(nz-1)*(ny-1)*(nx-1)
  ...
enddo
```

Fig. 7 Pseudo code of loop fusion

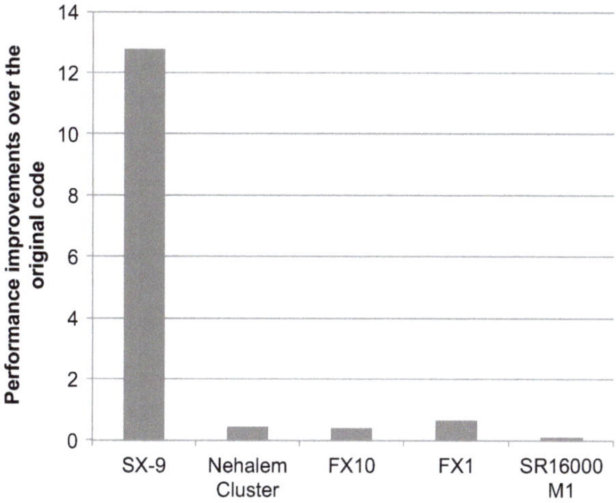

Fig. 8 Performance impacts of loop fusion

3.3 Ongoing and Future Work

As described in the previous subsections, now we accumulate 22 optimization cases, and four cases are detected as system-specific optimizations. By continuing these activities, we believe more cases which deteriorate the performance portability would be detected. Since analyzed applications are mainly optimized for vector supercomputers and limited to inner-node optimizations, we need to broaden our target applications to expand the contents of the optimization lists and the HPC refactoring catalog. In addition, aiming at supporting a legacy code migration to the future HPC system, the HPC refactoring catalog will be provided to Xevolver framework [5] as a transformation receipt in XSLT format. Furthermore, now we are preparing a web page to share our research activities with HPC communities.

Conclusions
While conventional refactoring has potentials to improve the design of the code in terms of readability and portability, HPC refactoring is introduced to realize performance portability among the various future HPC systems. In this paper, we describe our activities for designing the HPC refactoring catalog which describe the guidelines for keeping a high performance portability of HPC applications. Through a quantitative analysis and evaluations of the practical codes, the system specific optimizations are detected and accumulated. Now we expand the contents of the catalog, and it would be open for public in the near future.

Acknowledgements This research was partially supported by Core Research of Evolutional Science and Technology of Japan Science and Technology Agency (JST CREST) "An Evolutionary Approach to Construction of a Software Development Environment for Massively-Parallel Heterogeneous Systems." This research uses the SX-9 and Express5800 systems in Cyberscience Center of Tohoku University, the Hitachi SR16000M1 system in the Information Initiative Center of Hokkaido University, the Fujitsu PRIMEHPC FX1 and FX10 Systems in the Information Technology Center of Nagoya University.

References

1. Top500list. http://www.top500.org
2. Kjolstad, F., Dig, D., Snir, M.: Bringing the HPC programmer's IDE into the 21st century through refactoring. In: Proceedings of SPLASH 2010 Workshop on Concurrency for the Application Programmer (CAP'10) (2010)
3. Fowler, M., et al.: Refactoring: improving the design of existing code. Addison-Wesley, Reading (1999)
4. Momose, S., Hagiwara, T., Isobe, Y., Takahara, H.: The brand-new vector supercomputer, SX-ACE. In: Proceedings of ISC '14, pp. 194–214 (2014)
5. Takizawa, H., Egawa, R., Takahashi, D., Suda, R.: HPC refactoring with hierarchical abstractions to help software evolution. In: Sustained Simulation Performance 2012, pp. 27–33 (2013)

Endorsing Supercomputing Applications to Java Language

Alexey Cheptsov and Bastian Koller

Abstract We present a scalable, open source realization of MPI-2 standard for Java, seamlessly integrated in Open MPI, as a reaction to novel challenges of supercomputing in the web application domain. A number of Java software solutions developed for the Web, such as coming from Information Retrieval, Semantic Web, and other domains, have begun to face performance and scalability challenges, for which MPI has proved to be an efficient solution in the "traditional" high performance computing languages, such as C and Fortran. We demonstrate that the native Java language design prevents MPI implementations to scale massively on productional supercomputing systems, and present a solution of overcoming the scalability issues by integrating in the native C realization of Open MPI. We also point out the design features of Open MPI that enable the proliferation of MPI into Java applications. Finally, we present some successful pilot scenarios implemented with MPI in Java and discuss future work in terms of promising Java applications of Open MPI, such as Random Indexing of large semantically annotated text sets.

1 Introduction

The Web in its current status serves a large testbed infrastructure for scientific and engineering computing, whereby the Java programming language enjoys a dominant position [1, 2]. Driven by the concepts of portability and interoperability, Java has become a widely accepted general-purpose programming language, with a large existing code base and a diversity of programmer communities. Among others, Java has gained a wide adoption in web- and data-centric computing. Nowadays, processing exploding volumes of data exposed on the Web is the primary challenge for many areas of science and technology [3]. Some rich data-centric web domains, such as Information Retrieval, Resource Discovery, and Semantic Web have already shown a potential demand for parallel and high-performance computing adoption [4]. Whereas the recent advances of those communities require their respective Java applications to scale up to the requirements of the vast and continuously

A. Cheptsov (✉) • B. Koller
High Performance Computing Center Stuttgart, Nobelstr. 19, 70569 Stuttgart, Germany
e-mail: cheptsov@hlrs.de; koller@hlrs.de

M.M. Resch et al. (eds.), *Sustained Simulation Performance 2014*,
DOI 10.1007/978-3-319-10626-7_9

increasing data volumes, coming from hand-held devices to data centers or collected in extra-scale knowledge bases, such as Open PHACTS[1] or Linked Life Data,[2] the Java language, utilized mostly in the web domain, fairly lacks parallelization mechanisms that would enable improve performance and efficiency for data-centric web applications, ported to a modern supercomputing system.

At present, we observe a large shift of interests in the Web in general and Java in particular towards the exploitation of the supercomputing facilities. A number of the major software vendors in the Semantic Web domain have already established a collaboration with the high performance computing centers such as HLRS[3] in Stuttgart (Germany) or SARA[4] in Amsterdam (the Netherlands), and this trend is going to grow in the nearest future. Both commodity and more dedicated high performance computing (HPC) architectures, such as the Cray XMT [5], have been in focus of the data-intensive web applications. The latter, however, have proved success only for a limited number of tasks so far, which is mainly due to the complexity of adopting the offered software frameworks (mainly non-standard pragma-based language extensions) by the web applications.

In terms of the parallel software frameworks for the commodity HPC architectures, there are already some developments, such as Hadoop, which is a Java based and data-centric parallel development framework based on MapReduce[6], or Ibis [7], which is a programming environment for running parallel Java applications in a thread-fashioned way. Those developments prove that the web applications implemented in Java are able to successfully exploit HPC in order to scale to the sizes unreachable for native, thread-based implementations. However, none of those frameworks is near as efficient or well-developed as found in the Message-Passing Interface (MPI)[8]. MPI has become a de-facto standard in the area of parallel computing for C, C++, and Fortran applications. MPI allows an application to employ data and/or functional decomposition in order to split up its data- and/or workflow in separate processes, running in parallel and non-concurrently on distributed compute nodes of a supercomputing system. MPI also serves a communication layer among those nodes to synchronize the process execution in time as well as to exchange the data. The latter is done by means of the messages passed between (the point-to-point) or among (the collective) communications.

Given the vast problem sizes addressed by the modern web applications, and given the emergence of the new communities interested in adopting MPI, it seems natural to explore the benefits of MPI for Java applications on the HPC platforms as well. Introducing MPI for Java applications poses several challenges. First, the API set should be compliant with the MPI standard[8], but not downgrade

[1] http://www.openphacts.org

[2] http://www.linkedlifedata.org

[3] https://www.hlrs.de

[4] https://www.sara.nl/

the flexibility of the native Java language constructions. Second, the hardware support should be offered in a way that overcomes the limitation of the Java run-time environment, but provides such important features as thread-safety. Third, MPI support should be seamlessly integrated in the parallel application's execution environment. These three issues of functionality, adaptivity, and usability must be addressed in complex to make the use of MPI in Java applications practical and useful.

We look how to resolve these issues in a way that leverages the advances of the existing MPI frameworks. We present and evaluate our solution for introducing Java support in Open MPI [9], which is one of the most popular open source MPI-2 standard's implementations nowadays. Our approach is based on the integration of Java MPI bindings developed for mpiJava [10] directly in the native C realization of Open MPI, thus minimizing the bindings overhead and leveraging the Open MPI's run-time and development environment to ensure the high scalability of the Java parallel application. We also give examples of successful pilot scenarios implemented with our solution and discuss future work in terms of the development, implementation, and standardization activities.

2 Related Work

2.1 MPI Bindings for Java

Although the official MPI standard's declaration is limited to the C and Fortran languages only, there has been a number of standardization efforts made towards introducing the MPI bindings for Java. The most developed API set, however, has been proposed by mpiJava [11, 12] developers.

There are only a few approaches to implement MPI bindings for Java. These approaches can be classified in two following categories:

- Pure Java implementations, e.g. based on RMI (Remote Method Invocation) [13], which allows Java objects residing in different virtual machines to communicate with each other, or lower-level Java sockets API.
- Wrapped implementation using the native methods implemented in C languages, which are presumably more efficient in terms of performance as the code managed by the Java run-time environment.

In practice, none of the above-mentioned approaches satisfies the contradictory requirements of web users on application portability and efficiency. Whereas a pure Java implementation, e.g. MPJ Express [14] and MPJ/Ibis [7] do not benefit from the high speed interconnects, such as Infiniband[15] and thus introduces communication bottlenecks and does not demonstrate acceptable performance on the majority of today's production HPC systems, a wrapped implementation, e.g. mpiJava requires a native C library, which can cause integration and interoperability issues with the underlying MPI implementation.

In looking for a trade-off between the performance and the usability, and also in view of the complexity of realizing Java support for high speed cluster interconnects [16], the most promising solution seems providing Java support directly in the native MPI implementation in C.

2.2 Native C Implementations of MPI

Despite a great variety of the native MPI implementations, there are only a few of them that address the requirements of Java parallel applications on process control, resource management, latency awareness and management, and fault tolerance. Among the sustainable open-source implementations, the most related to our goals to implement the Java MPI bindings are Open MPI[5] and MPICH2.[6] Both Open MPI and MPICH2 are open-source, production quality, and widely portable implementation of the MPI-1 and MPI-2 standards. Although both libraries claim to provide a modular and easy-to-extend framework, the software stack of Open MPI seems to better suit the goal of introducing a new language's bindings, which our research aims to. The architecture of Open MPI [9] defines a dedicated layer used to introduce bindings, which are currently provided for C, F77, F90 and some other languages. Extending the MPI-Layer with the Java language support seems to be a very promising implementation of the proposed integration of Java bindings, taking benefits of all the layers composing Open MPI's architecture. The only problem with the Open MPI implementation seems at the moment to be some luck of the modern Cray's interconnect (Gemini[7]) support. However, we are aware of several oncoming actions in the Open MPI community performed in this direction, so we expect this feature to be supported in one of the upcoming official releases.

2.3 Non-MPI Based Approaches

There are only a few alternatives to MPI in introducing the large-scale parallelism to Java applications. The most promising among those alternatives in terms of the performance and usability are solutions offered by IBIS/JavaGAT and MapReduce/Hadoop.

IBIS [17] is a middleware stack used for running Java applications in distributed and heterogeneous computing environments. IBIS leverages the peer-to-peer communication technology by means of the proprietary RMI implementation, based on GAT (Grid Access Toolkit). The Java realization of GAT (JavaGAT [18]) is a

---

[5]http://www.open-mpi.org

[6]http://www.mcs.anl.gov/research/projects/mpich2/

[7]http://www.cray.com/Products/XE/Technology.aspx

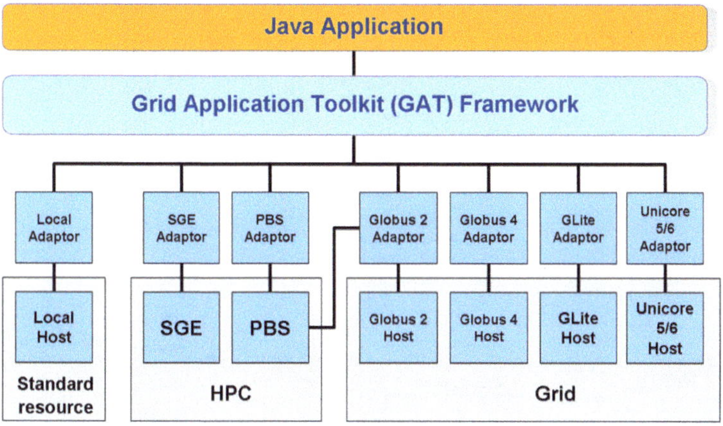

Fig. 1 Remote Java method invocation with JavaGAT

middleware stack that allows the Java application to instatiate its classes remotely on the network-connected resource, as depicted in Fig. 1. Along with the traditional access protocols. e.g. telnet or ssh, the advanced access methods, such as ssh-pbs for clusters or gsissh for grid infrastructures are supported. IBIS implements a mechanism of multiple fork-joins to detect and decompose the application's workload and execute its parts concurrently on distributed resources. While [7] indicates some successful Java applications implemented with IBIS/JavaGAT and shows a good performance, there is no clear evidence about the scalability of this solution for more complex communication patterns, involving nested loops or multiple split-joins. Whereas IBIS serves a very effective solution for the distributed computing environments, e.g. Grid or Cloud, it's definitively not the best approach to be utilized on the tightly-coupled productional clusters.

MapReduce framework [6] and its most prominent implementation in Java, Hadoop, has gained tremendous popularity in modern data-intensive application scenarios. MapReduce is a programming model for data-centric applications exploiting large-scale data parallelism, originally introduced by Google in its search engine. In MapReduce, the application's workflow is divided into three main stages (see Fig. 2): map, process, and reduce. In the map stage, the input data set is split into independent chunks and each of the chunks is assigned to the independent tasks, which are then processed in the completely parallel manner (process stage). In the reduce stage, the output produced by every map task is collected, combined and the consolidated final output in produced.

The Hadoop framework is a service-based implementation of MapReduce for Java. Hadoop considers a parallel system as a set of master and slave nodes, deploying on them services for scheduling tasks as jobs (Job Tracker), monitoring the jobs (Task Tracker), managing the input and output data (Data Node), re-executing the failed tasks, etc. This is done in a way that ensures a very high service reliability and fault tolerance properties of the parallel execution. In Hadoop,

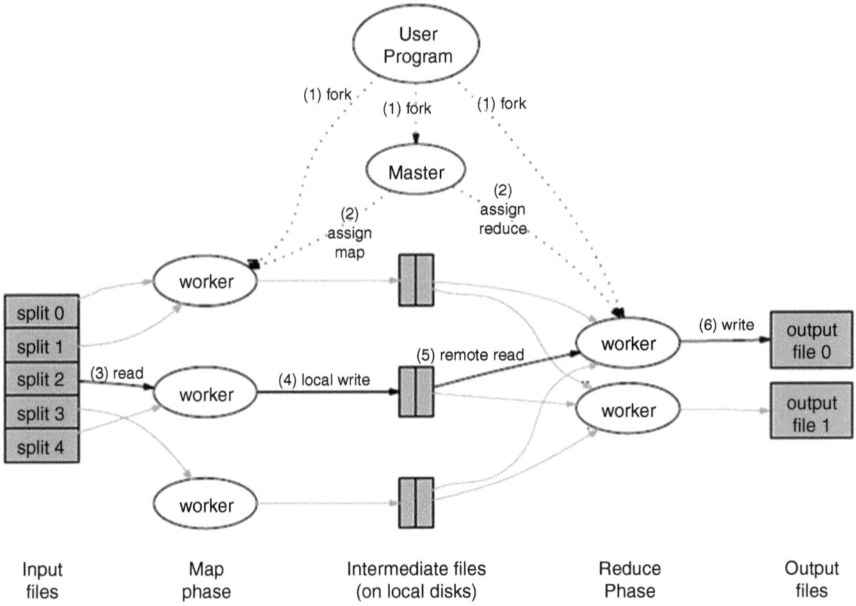

Fig. 2 MapReduce processing schema

both the input and the output of the job are stored in a special distributed file-system. In order to improve the reliability, the file also provides an automatic replication procedure, which however introduces an additional overhead to the inter-node communication. Due to this overhead, Hadoop ensures much more poor performance than MPI, however offering better QoS characteristics related to the reliability and fault-tolerance. Owing to the fact that both MPI and MapReduce paradigms have been designed to serve different purposes, it is hardly possible to comprehensively compare them. However they will obviously benefit from a cross-fertilization; for example, Open MPI can serve a high-performance communication layer to Hadoop that would allow for omitting the usage of the file system for distributing the map and gathering the reduce tasks.

3 Design and Implementation

3.1 Objectives

We have based our Java MPI bindings on the *mpiJava* code, originally developed in HPJava[8] project and currently maintained by the High Performance Computing

[8]http://www.hpjava.org

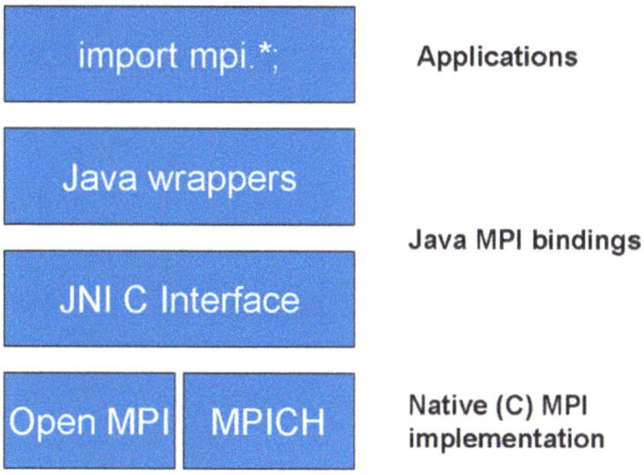

Fig. 3 MpiJava architecture

Center Stuttgart.[9] MpiJava provides a set of Java Native Interface (JNI) wrappers to the native MPI-1.1 communication methods, as shown in Fig. 3. JNI enables the programs running inside a Java run-time environment to invoke native C code and thus use platform-specific features and libraries [19], e.g. the Infiniband software stack. The application-level API is constituted by a set of Java classes, designed in conformance to the MPI-1.1 standard and the specification in [11]. The Java methods internally invoke the MPI-C functions using the JNI stubs. The realization details for mpiJava can be obtained from [10, 20].

The major integration tasks were as follows:

- extend the Open MPI architecture to support Java bindings
- extend the previously available mpiJava bindings to MPI-2 (and possibly MPI-3) standard
- improve the native Open MPI configuration, build, and execution system to seamlessly support the Java bindings
- redesign the Java interfaces that use JNI in order to better confirm to the native realization
- optimize the JNI code to minimize its invocation overhead based on the ones acknowledged in the MPI community
- create test applications for performance benchmarking

Below we discuss these issues in a new Java bindings design, incorporated into the Open MPI development branch.

3.2 Architecture

Open MPI is a high performance, production quality, MPI-2 standard compliant implementation. Open MPI consist of three abstraction layers that combine to

[9] http://sourceforge.net/projects/mpijava/

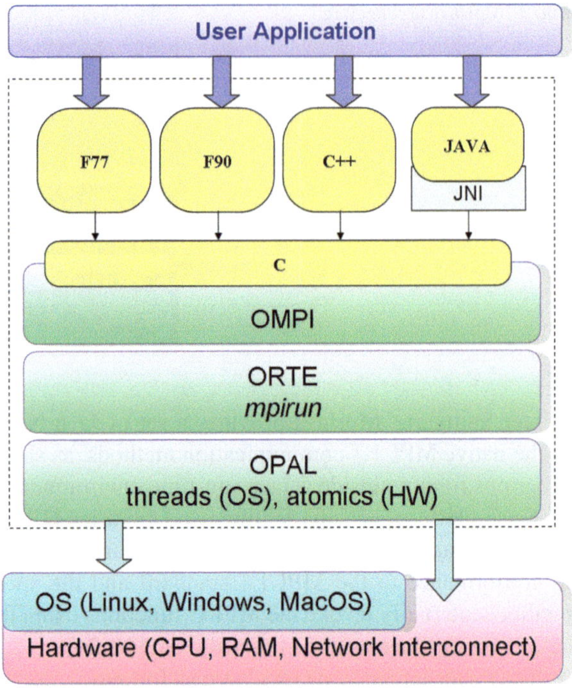

Fig. 4 Open MPI architecture

provide a full featured MPI implementation: Open Portable Access Layer (OPAL) that abstracts the peculiarities of a specific system away to provide a consistent interface aiding portability, Open Run-Time Environment (ORTE) that provides a uniform parallel run-time interface regardless of system capabilities, and Open MPI (OMPI) that presents the application with the expected MPI standard interface. Figure 4 shows the Open MPI architecture, enhanced with the Java bindings support.

Both Java classes and JNI code for calling the native methods were integrated into Open MPI. The major integration effort required however the OMPI (Java classes, JNI code) and the ORTE (run-time specific options) levels. The implementation of the Java class collection followed the same strategy as for the C++ binding realization, for which the opaque C objects are encapsulated into suitable class hierarchies and most of the library functions are defined as class member methods. The architecture of Java classes integrated into OMPI is shown in Fig. 5. Along with the classes implementing the MPI-1.2 functionality (MPI package), the collection includes the classes for error handling (Errhandler, MPIException), datatypes (Datatype), communicators (Comm), etc. More information about the implementation of both Java classes and JNI-C stubs can be found in previous publications [10, 15].

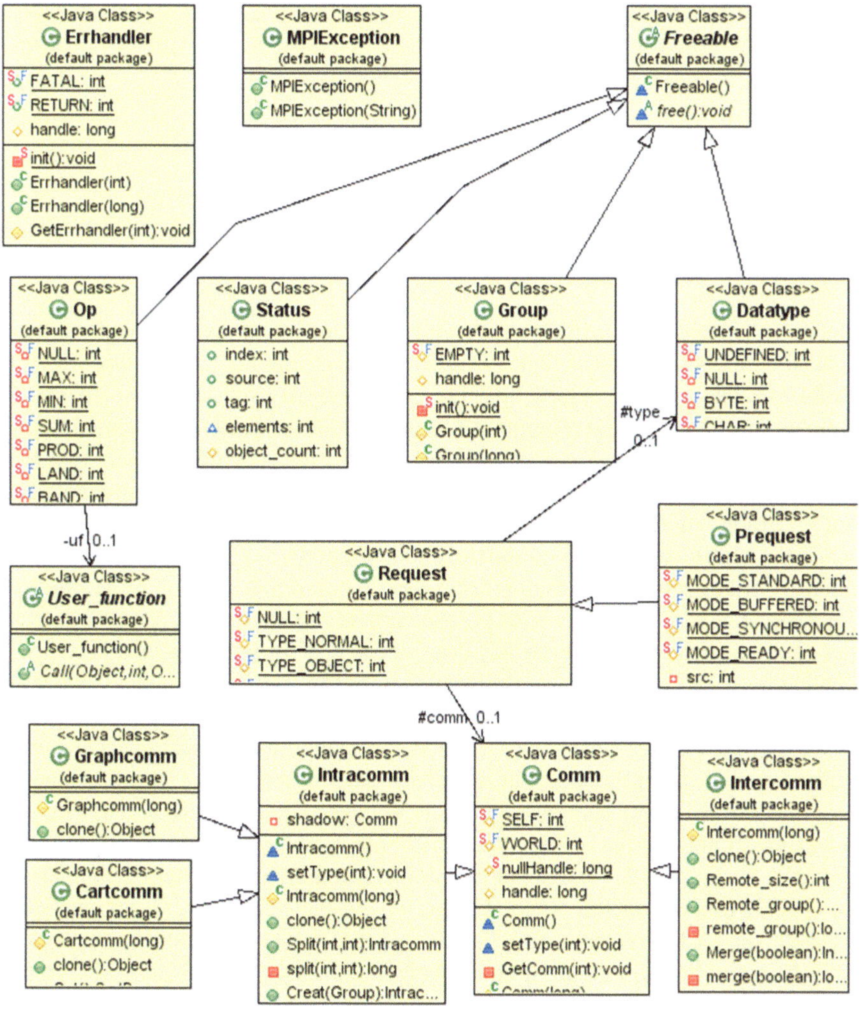

Fig. 5 ompiJava class collection

3.3 *Configuration and Running*

A crucial aspect of the Java binding integration into Open MPI was setting up a common configuration and build system. The original mpiJava installation procedure is very restrictive with regard to Open MPI and required a lot of adaptation, being thus of a big challenge for an unexperienced user. For example, Java bindings required Open MPI to be built at least with shared libraries (i.e., *–enable-shared*). Moreover, several object-level conflicts had to be resolved for the JNI code on the linking stage. We enhanced the native (autotools based) Open MPI configuration

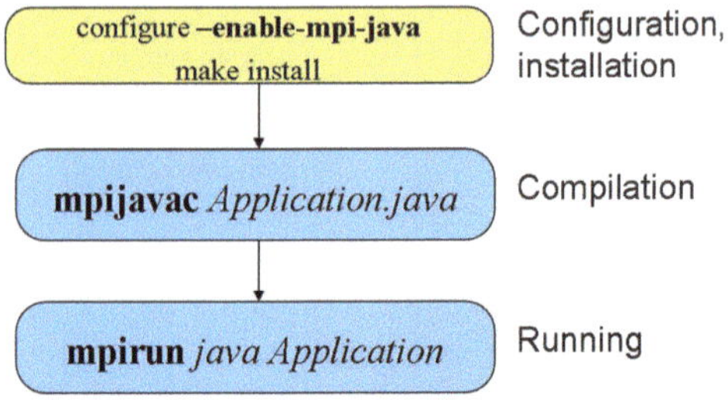

Fig. 6 Schema of ompiJava application running

and build system to extensively support the Java bindings. In particular, there was a special configuration instruction (*–enable-mpi-java*) implemented in Open MPI to enable producing the Java bindings. The configuration script automatically checks whether the JDK (Java Development Kit) is provided in the default location, unless the options *–with-jdk-bindir* and *–with-jdk-headers* are specified to point to a non-standard location. The Java bindings will build only if they are explicitly requested and the required JDK (Java Development Kit) support is found. If the *–enable-mpi-java* option is switched on during the configuration, a user-level Java library, *mpi.jar*, will be produced, which is necessary for compiling and running the parallel Java application. By analogy with the standard compilers for C, C++, and Fortran, the *mpijavac* wrapper compiler has been developed for building Java-based MPI applications. The wrapper ensures that all required MPI libraries and class paths are defined.

To support the user during the execution of parallel Java applications, the *mpirun* script has also been extended to detect the java command and ensure that the required MPI libraries and class paths are defined to support execution. The user therefore do not need to specify the Java library path to the MPI installation, nor the MPI classpath. Any classpath definitions required for the application should be specified either on the command line or via the *CLASSPATH* environmental variable; the local directory will be added to the class path if nothing is specified by default. The schema in Fig. 6 summarizes the main steps towards getting a running MPI-parallel Java application, such the simple application in Listing 1.

```
import java.io.*;
import mpi.*;

class Hello {
  public static void main(String[] args) throws MPIException
  {

    int my_pe, npes;

    MPI.Init(args);

    my_pe = MPI.COMM_WORLD.Rank();
    npes  = MPI.COMM_WORLD.Size();
```

```
        System.out.println("Hello from " + my_pe +
            " out of " + npes);

        MPI.Finalize();
    }
}
```

Listing 1 Simple Java application using MPI

4 Performance Evaluation

We evaluated our implementation of the Java bindings in Open MPI on up to 128 compute nodes of the NEC Nehalem[10] cluster of HLRS (Intel Xeon CPU, Infiniband interconnect). First, we performed our evaluation on the standard MPI benchmark set, including point-to-point (ping-pong) and global collective (gather) communication tests. Then, we evaluated the performance of a pilot Java application that performs Random Indexing of large RDF text sets.

4.1 Basic Benchmarks

We examined two types of communication: point-to-point (between two nodes) and collective (between a group of nodes), varying the size of the transmitted messages. We prepared a set of benchmarks in Java based on those well-recognized in the MPI community, e.g. NetPIPE[11] or NAS.[12] In order to evaluate the JNI code overhead, we reproduced the benchmarks also in C and run them with the native Open MPI. In order to increase the accuracy of the evaluation and possibly eliminate performance degradation issues, e.g. caused by the network interconnect and CPU side-effects, we implemented a sampling mechanism that makes $N = 16$ blank communications before the actual measurement happens. Moreover, every communication is done $M = 16$ times and the normalized time measured. The time (wall clock) is measured using *WTime*. The source code of the used benchmarks as well as exact timings can be found uploaded at SourceForge.[13]

---

[10]http://www.hlrs.de/systems/platforms/nec-nehalem-cluster/

[11]http://www.scl.ameslab.gov/netpipe/

[12]http://www.nas.nasa.gov/publications/npb.html

[13]http://sourceforge.net/projects/mpijava/

We ran all benchmarks with the following MPI implementations:

- **ompiC**—native C implementation of Open MPI (the actual trunk version), build with the GNU compiler (v.4.6.1),
- **ompiJava**—our implementation of Java bindings on top of *ompiC*, running with Java JDK (v.1.6.0), and
- **mpj**—the newest version of MPJ-Express (v.0.38), a native implementation, running with the same version of Java.

We did intentionally not rely on the previously reported benchmarks[14] in order to eliminate the measurement deviations that might be caused by running tests in a different hardware or software environment. Moreover, in order to ensure a fair comparison between all these three implementations, we ran each test on the absolutely same set of compute nodes.

The point-to-point benchmark implements a "ping-pong" based communication between two single nodes; each node exchanges the messages of growing sizes with the other by means of blocking Send and Receive operations. Since Open MPI uses different protocols for transmitting small (up to 1,024 KBytes) and large (more than 1,024 KBytes) messages, we perform a separate evaluation for both those size ranges. In both cases (Figs. 7 and 8), our *ompiJava* implementation was as expected not as efficient as the underlying *ompiC*, but showed much better performance as the socket-based *mpj*. Regardless of the message size, *ompiJava* achieves around eight times higher throughput as *mpj* has demonstrated (Fig. 9). The chart with the message rate comparison between ompiJava and mpj is shown in Fig. 10. In addition, we examined the communication between the processes resided on a single compute node, i.e. inter-node communication. Figure 11 demonstrates that all three implementations take advantages of the shared memory communication protocols

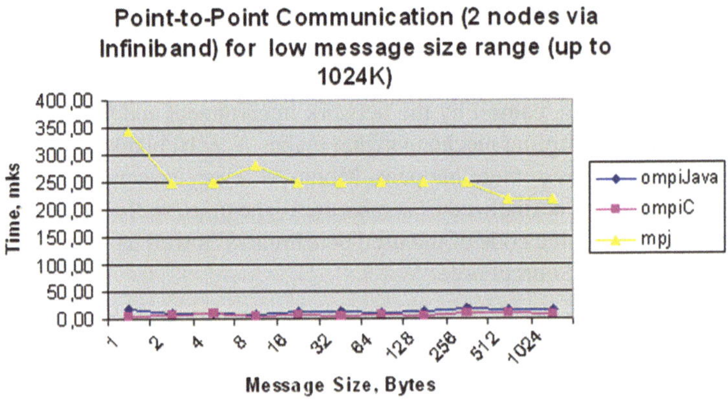

Fig. 7 Point-to-point communication performance for the low message size range

[14]http://mpj-express.org/performance.html

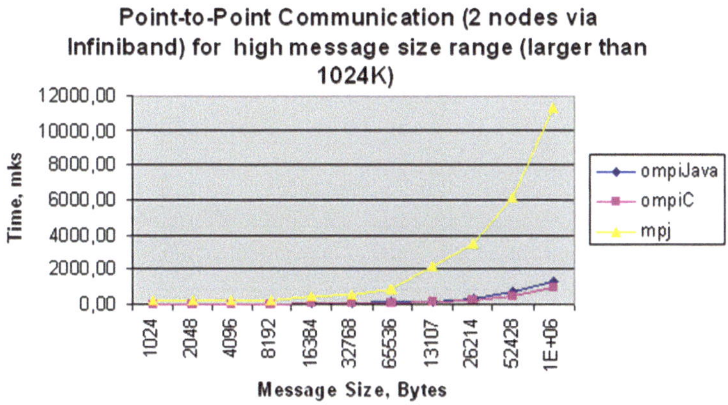

Fig. 8 Point-to-point communication performance for the high message size range

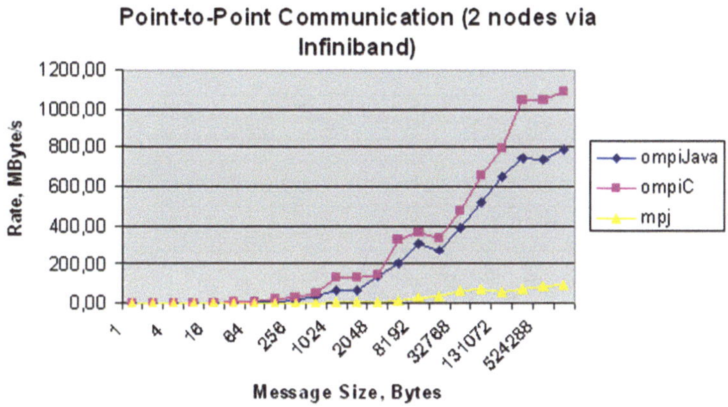

Fig. 9 Message rate for the point-to-point communication (for both low and high message size ranges)

and that the performance of both Java based implementations, ompiJava and mpj, is pretty comparable for the large messages. However, ompiJava was around four times more efficient on the small message size range (up to 1,024 KByte).

The collective communication benchmark implements a single blocking message gather from all the involved nodes. Figure 12 shows the results collected for $P = 2^k$(where $k = 2-7$) nodes, with a varying size of the gathered messages. The maximal size of the aggregated data was 8 GByte on 128 nodes. Figure 13 demonstrates the comparison of collective gather performance for all tested implementations on the maximal number of the available compute nodes (128). Whereas the Infiniband-aware *ompiJava* and *ompiC* scaled quite well, the TCP based *mpj* has shown very poor performance; for the worst case (on 128 nodes) a slow-down up to 30 times compared with *ompiJava* was observed.

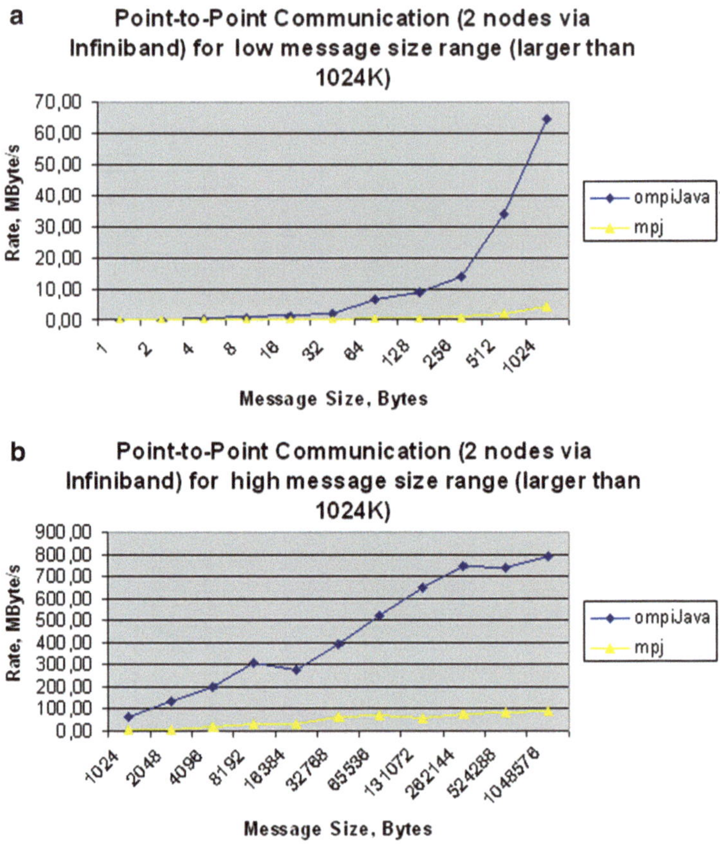

Fig. 10 Comparison of the message rate for ompiJava and mpj for (**a**) low and (**b**) high message size range

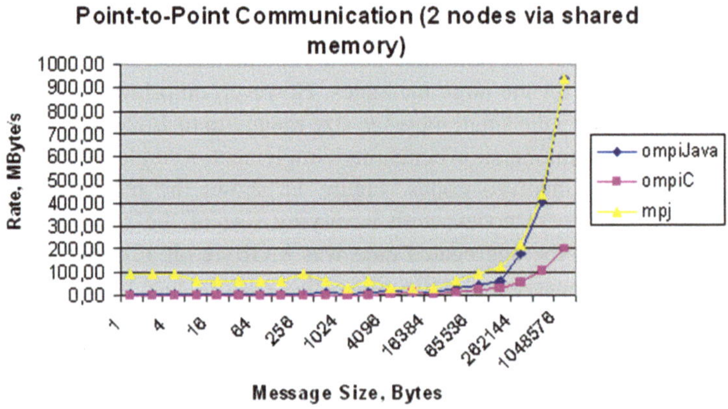

Fig. 11 Message rate for the inter-node point-to-point communication

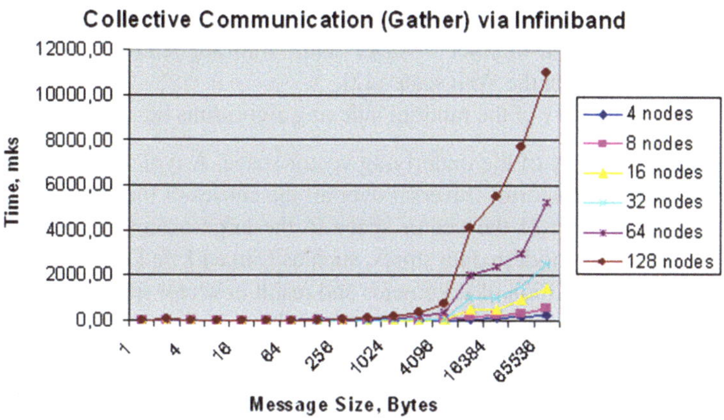

Fig. 12 Collective gather communication performance of ompiJava

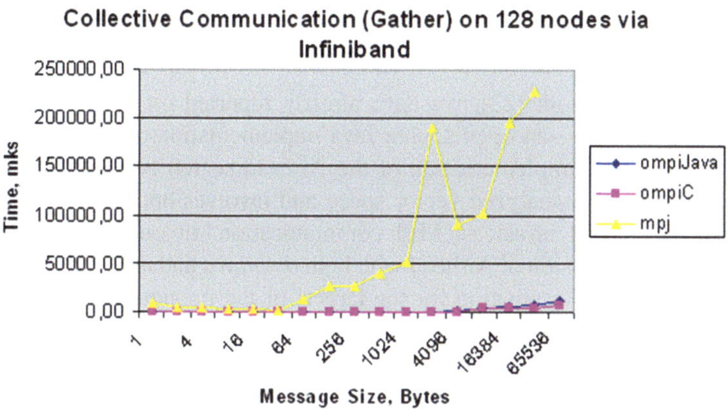

Fig. 13 Collective gather communication performance on 128 nodes

All the timings produced by our benchmarks can be found uploaded at SourceForge.[15]

4.2 *Pilot Application Scenario: Random Indexing Over Large Text Sets*

Random indexing [21] is a distributional statistic technique used in resource discovery for extracting semantically similar words from the word co-occurrence statistics

[15]http://sourceforge.net/projects/mpijava/

in the text data, based on high-dimensional vector spaces. Random indexing offers new opportunities for a number of large-scale Web applications performing the search and reasoning on the Web scale [22].

The main complexity of the random indexing algorithms lies in the following:

- High dimensionality of the underlying vector space. A typical random indexing search algorithm performs traversal over all the entries of the vector space. This means, that the size of the vector space to the large extent defines the search performance. The modern data stores, such as Linked Life Data or Open Phacts consolidate many billion of statements and result in vector spaces of a very large dimensionality. Random indexing over such large data sets is computationally very costly, with regard to both execution time and memory consumption. The latter is of especial drawback for use of random indexing packages on the mass computers. So far, only relatively small parts of the Semantic Web data have been indexed and analyzed.
- High call frequency. Both indexing and search over the vector space is typically a one-time operation, which means that the entire process should be repeated from scratch every time new data is encountered.

In our previous work [23], we have already reported on the efforts done on parallelizing Airhead—an open source Java implementation of Random Indexing algorithm. Our MPI implementation of the Airhead search is based on a domain decomposition of the analyzed vector space and involves both point-to-point and collective gather and broadcast MPI communication. In our current work, we evaluated the MPI version of Airhead with both *ompijava* and *mpj* implementations.

We performed the evaluation for the largest of the previously reported in [23] data sets (Wiki2) that comprises 1 Million of high density documents and occupies 16 GByte disk storage space. The overall execution time (wall clock) was measured. Figure 14a shows that both *ompijava* and *mpj* scale up well until the problem size is large enough to saturate the capacities of a single node. Nevertheless, our implementation was around 10% more efficient over *mpj* (Fig. 14b).

5 Future Work

Our future work is going to concentrate on development, dissemination, standardization, and support actions around ompiJava.

Development will mainly focus on extending the Java binding to fully support the MPI-3 specification. On the Open MPI side, the configuration and run-time environment will be improved as well to provide better support to Java applications.

Standardization will aim at introducing Java language-specific bindings into the MPI standard. Adding the Java API specification to the MPI standard is appropriate to the novel trends in supercomputing. Although there have been several standardization activities seeking to come up with a common specification of MPI interface for Java, such as Message-Passing Working Group of the Java Grande

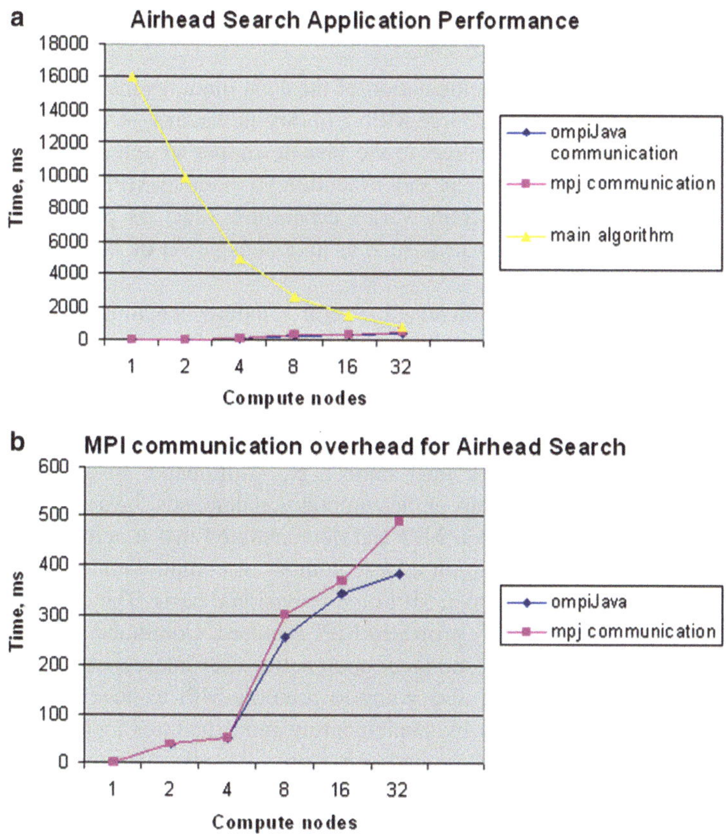

Fig. 14 Airhead search application performance: a) overall execution time, b) communication overhead time

Forum,[16] there has not been a lot of results impacted the MPI Forum so far and we hope to change this situation.

The main goal of the dissemination activities is to promote MPI to the web application communities owning data-intensive use cases, such as Statistical Semantics and Reasoning[24].

Introducing new tools supporting the developers during performance analysis and debugging of the parallel Java applications, such as Eclipse Parallel Tool Platform,[17] is going to be the main focus of the supporting activities.

[16]http://www.javagrande.org

[17]http://www.eclipse.org/ptp/

Conclusion

We introduced a new implementation of the Java bindings integrated in one of the most popular open source MPI-2 library nowadays—Open MPI. The integration allowed us to deliver to the Java developer an extensive software environment for development and execution of parallel-MPI applications, leveraging the Open MPI framework's capabilities, such as portability and usability, with those of mpiJava, such as an extensive set of Java-based API for MPI communication.

We implemented and conducted several standard benchmarks revealing that the new implementation allows a Java application to achieve an acceptable performance on a commodity cluster. We proved that mpiJava's JNI implementation although introduces an overhead to the native C methods, but outperforms the alternative "native Java" implementation on the cluster with a high-speed network interconnect, e.g. Infiniband. In addition, we ported a challenging, in terms of the computation demands, Java applications, Airhead Search, to the Open MPI and demonstrated that it achieves a very good performance speed-up on a large number of compute nodes. The Java bindings are provided in Open MPI on a provisional basis. They are still not a part of the current or any proposed MPI standard. Continued inclusion of the Java bindings will be contingent upon active user interest and continued developer support. We are also going to promote MPI to new, challenging data-intensive applications by implementing more features and developing tools for performance analysis and debugging. We also plan to collaborate with Hadoop on investigation of opportunities for MPI to serve the Hadoop's communication layer.

Acknowledgements We thank Open MPI consortium for the support with porting mpiJava bindings.

References

1. Alexandrov, A., Ibel, M., Schauser, K., Scheiman, C.: SuperWeb: research issues in Java-based global computing. In: Proceedings of the Workshop on Java for High performance Scientific and Engineering Computing Simulation and Modelling. Syracuse University, New York
2. Fox, G., Furmanski, W.: Java for parallel computing and as a general language for scientific and engineering simulation and modeling. Concurr. Comput. **9**(6), 415–425 (1997)
3. McKendrick, J.: Size of the data universe: 1.2 zettabytes and growing fast. ZDNet report (2010)
4. Cheptsov, A., Assel, M.: Towards high performance semantic web – experience of the LarKC project. J. Innov. Supercomput. Dtschl. **9**(1), 72–75 (2011)
5. Goodman, E., Jimenez, E., Mizell, D., al-Saffar, S., Adolf, B., Haglin, D.: High-performance computing applied to semantic databases. In: Proceedings of the Extended Semantic Web Conference (ESWC), vol. 2, pp. 31–45 (2011)

6. Dean, J., Ghemawat, S.: MapReduce- simplified data processing on large clusters. Retrieved from Google Library (2004)
7. Bornemann, M., van Nieuwpoort, R., Kielmann, T.: MPJ/Ibis: a flexible and efficient message passing platform for java. In: Proceedings of 12th European PVM/MPI Users' Group Meeting, pp. 217–224 (2005)
8. MPI: A Message-Passing Interface Standard. Message Passing Interface Forum. http://www.mcs.anl.gov/research/projects/mpi/mpi-standard/mpi-report-1.1/mpi-report.htm
9. Gabriel, E., Fagg, G., Bosilca, G., Angskun, T., Dongarra, J., Squyres, J., Sahay, V., Kambadur, P., Barrett, B., Lumsdaine, A., Castain, R., Daniel, D., Graham, R., Woodall, T.: Open MPI: goals, concept, and design of a next generation MPI implementation. In: Proceedings, 11th European PVM/MPI Users' Group Meeting, Budapest, September 2004
10. Baker, M., Carpenter, B., Fox, G., Ko, S., Lim, S.: mpiJava: an object-oriented java interface to MPI. International Workshop on Java for Parallel and Distributed Computing, IPPS/SPDP 1999, San Juan, April 1999
11. Carpenter, B., Fox, G., Ko, S., Lim, S.: mpiJava 1.2: API specification. http://citeseer.ist.psu.edu/ko99mpijava.html (1999)
12. Carpenter, B., Getov, V., Judd, G. Skjellum, A., Fox, G.: MPJ: MPI-like message passing for Java. Concurr. Comput. **12**(11), 1019–1038 (2000)
13. Kielmann, T., Hatcher, P., Bougé, L., Bal, H.: Enabling java for high-performance computing: exploiting distributed shared memory and remote method invocation. Commun. ACM (2011)
14. Baker, M., Carpenter, B., Shafi, A.: MPJ express: towards thread safe java HPC. In: Proceedings of the IEEE International Conference on Cluster Computing (Cluster 2006), Barcelona, Spain, 25–28 September 2006. Retrieved from http://mpj-express.org/docs/papers/mpj-clust06.pdf (2006)
15. Judd, G., Clement, M., Snell, Q., Getov, V.: Design issues for efficient implementation of MPI in java. In: Proceedings of the 1999 ACM Java Grande Conference, pp. 58–65 (1999)
16. Zhang, H., Huang1, W., Han, J., He, J., Zhang, L.: A performance study of java communication stacks over InfiniBand and gigabit ethernet. In: 2007 IFIP International Conference on Network and Parallel Computing - Workshops (2007)
17. van Nieuwpoort, R., Maassen, J., Wrzesinska, G., Hofman, R., Jacobs, C., Kielmann, T., Bal, H.: Ibis: a flexible and efficient Java based grid programming environment. Concurr. Comput. **17**(7–8), 1079–1107 (2005)
18. van Nieuwpoort, R., Kielmann, T., Bal, H.: User-friendly and reliable grid computing based on imperfect middleware. In: Proceedings of the ACM/IEEE Conference on Supercomputing (SC'07), November 2007
19. Liang, S.: Java native interface: programmer's guide and reference. Addison-Wesley, Boston (1999)
20. Vodel, M., Sauppe, M., Hardt, W.: Parallel high-performance applications with MPI2Java - a capable java interface for MPI 2.0 libraries. In: 16th Asia-Pacific Conference on Communications (APCC), pp. 509–513 (2010)
21. Sahlgren, M.: An introduction to random indexing. In: Proceedings of Methods and Applications of Semantic Indexing Workshop at the 7th International Conference on Terminology and Knowledge Engineering TKE 2005, pp. 1–9 (2005)
22. Jurgens, D., Stevens, K.: The S-space package: an open source package for word space models. In: Proceedings of the ACL 2010 System Demonstrations, pp. 30–35 (2010)
23. Assel, M., Cheptsov, A., Czink, B., Damljanovic, D., Quesada, J.: MPI realization of high performance search for querying large RDF graphs using statistical semantics. In: Proceedings of the 1st Workshop on High-Performance Computing for the Semantic Web (HPCSW2011), co-located with the 8th Extended Semantic Web Conference, ESWC2011, Heraklion, 29 May 2011
24. Fensel, D., van Harmelen, F.: Unifying reasoning and search to web scale. IEEE Internet Comput. **11**(2), 96–95 (2007)

Performance Evaluation of an OpenMP Parallelization by Using Automatic Parallelization Information

Kazuhiko Komatsu, Ryusuke Egawa, Hiroyuki Takizawa, and Hiroaki Kobayashi

Abstract To exploit the potential of many core processors, a serial code is generally optimized for a particular compiler called a target compiler, so that the compiler can understand the code structure for automatic parallelization. However, the performance of such a serial code is always not portable to a new system that uses a different compiler. To improve the performance portability, this paper proposes an OpenMP parallelization method by using compiler messages of the target compiler. Since the compiler messages from the target compiler are also useful to identify key loop nests even for the different system, a programmer can use the message to easily parallelize a serial code with low programming effort. Furthermore, programmer's intention of the optimization can be migrated to other systems through the OpenMP parallelization, which results in high performance portability. The experimental results indicate that the OpenMP codes parallelized by the proposed method can achieve a comparable or even better performance than the automatically parallelized codes by various compilers.

1 Introduction

The number of cores on recent high-performance computing (HPC) systems has been drastically increased. From the top 500 lists [1], MilkyWay2, the top 1 system in November 2013, is equipped with more than 3 million cores, while

---

K. Komatsu (✉) • R. Egawa
Cyberscience Center, Tohoku University/JST CREST, 6-3 Aramaki-aza-aoba, Aoba, Sendai 980-8578, Japan
e-mail: komatsu@isc.tohoku.ac.jp; egawa@isc.tohoku.ac.jp

H. Takizawa
Graduate School of Information Sciences, Tohoku University/JST CREST, 6-6-01 Aramaki-aza-aoba, Aoba, Sendai 980-8579, Japan
e-mail: tacky@isc.tohoku.ac.jp

H. Kobayashi
Cyberscience Center, Tohoku University, 6-3 Aramaki-aza-aoba, Aoba, Sendai 980-8578, Japan
e-mail: koba@isc.tohoku.ac.jp

M.M. Resch et al. (eds.), *Sustained Simulation Performance 2014*,
DOI 10.1007/978-3-319-10626-7_10

Roadrunner, the top 1 system in November 2008, is equipped with only about 129 thousand cores. This drastic increase is caused by the trend to improve the computational capability. One of the main reasons is the increased number of cores in a processor. Even in a processor, 16 cores are integrated. An accelerator has more cores. For example, 61 cores are integrated into an Intel Xeon Phi. Thus, the parallelization to exploit computational capabilities of many cores becomes essential. A code has to be parallelized so as to use multiple cores by multiple threads, called *multithread parallelization*. In this paper, multithread parallelization using the OpenMP programming model is called *OpenMP parallelization*.

One of the most widely used approaches to multithread parallelization is to use automatic parallelization functions of compilers. One advantage of automatic parallelization is programmer's productivity. By compiling a code with automatic optimization options, a user can try and enjoy the computational capability of many cores without any code modifications. Another advantage is that a compiler can employ only appropriate parallelization for a target system by considering the characteristics of the target system[2]. Furthermore, since it is easy to use automatic parallelization functions, most of codes are modified by hand so that a particular compiler, called a *target compiler*, can easily understand the code structure, and analyze the data dependency. As a result, it is ensured that the target compiler can automatically parallelize the code and thus improve its performance.

However, the performance of the code is not always portable to other systems with different compilers, as automatic optimizations of a compiler are performed to exploit the potentials of the target system. To improve the performance portability of a code, OpenMP parallelization is one of the promising approaches. Although appropriate OpenMP directives need to explicitly be inserted to parallel regions of a code, OpenMP codes can be executed on any HPC system. This paper proposes an OpenMP parallelization method by using automatic parallelization information given by the target compiler. Since the information obtained from the compiler messages are often effective even for different systems as well as the target system, the proposed OpenMP parallelization method utilizes the information to help a user easily parallelize a serial code.

The rest of this paper is organized as follows. Section 2 proposes an OpenMP parallelization method that utilizes the automatic parallelization information of the target compiler. Section 3 evaluates the performances of the migrated OpenMP codes on various HPC systems. Finally, Sect. 4 gives conclusions of this paper.

```
$ sxf90 -Pauto -Chopt -R2 -Wf,-pvctl fullmsg,-L objlist,summary himeno.f90
...
mul(10):    himeno.f90, line 305: Parallel routine generated : jacobi$1
mul(1):     himeno.f90, line 305: Parallelized by PARDO.
vec(1):     himeno.f90, line 307: Vectorized loop.
vec(29):    himeno.f90, line 307: ADB is used for array.: a
vec(29):    himeno.f90, line 307: ADB is used for array.: bnd
vec(29):    himeno.f90, line 307: ADB is used for array.: wrk1
vec(29):    himeno.f90, line 307: ADB is used for array.: c
vec(29):    himeno.f90, line 307: ADB is used for array.: b
vec(29):    himeno.f90, line 307: ADB is used for array.: p
vec(29):    himeno.f90, line 307: ADB is used for array.: wrk2
mul(4):     himeno.f90, line 321: CRITICAL section.
vec(26):    himeno.f90, line 321: Macro operation Sum/InnerProd.
```

```
15:mul(10):    himeno.f90, line 326: Parallel routine generated : jacobi$2
16:mul(1):     himeno.f90, line 326: Parallelized by PARDO.
17:opt(1057):  himeno.f90, line 326: Complicated uses of variable inhibits loop
     optimization.
18:opt(1592):  himeno.f90, line 326: Outer loop unrolled inside inner loop.
19:vec(4):     himeno.f90, line 326: Vectorized array expression.
20:vec(3):     himeno.f90, line 326: Unvectorized loop.
21:vec(13):    himeno.f90, line 326: Overhead of loop division is too large.
22:vec(4):     himeno.f90, line 326: Vectorized array expression.
23:vec(29):    himeno.f90, line 326: ADB is used for array.: p
24:...
```

Listing 1 A compiler messages from the NEC SX compiler

2 OpenMP Parallelization by Using Automatic Parallelization Information

This paper proposes an OpenMP parallelization method by using the automatic parallelization information that helps a programmer find parallelizable loop nests and insert appropriate OpenMP directives. In general, an HPC code is optimized assuming a target compiler. As a result, the target compiler can easily detect and analyze the code structure and data dependencies necessary for multithread parallelization. The information about how the target compiler performs automatic optimizations is helpful to parallelize a serial code to the OpenMP model.

The compiler message contains which line in the code is parallelized, vectorized, and/or optimized, which line should be a critical section, and so on. Listing 1 shows an example of the compiler messages from the NEC SX compiler at compilation of the Himeno benchmark [3]. Lines 4 and 16 of the compiler message show the information of automatic parallelization. The message indicates that the loop nests from Lines 305 and 326 of *himeno.f90* can be automatically parallelized. Therefore, a programmer can easily understand that these loop nests can be parallelized by inserting appropriate OpenMP directives. As the parallelizable loop nests can be identified by reading the compiler messages instead of the code, the burden of the OpenMP parallelization can be drastically reduced.

The procedure of the proposed OpenMP parallelization method consists of the following steps.

1. Compile a serial code with enabling automatic parallelization functions to obtain compiler messages.
2. Identify parallelizable loop nests by checking the compiler messages. As the compile messages contain which parts of the code can be parallelized, parallelizable loop nests can easily be identified.
3. Find variables that need to be listed in the reduction clauses of OpenMP directives. If a loop nest includes a reduction operation on a variable, the variable is a so-called reduction variable. All reduction variables must be listed in the

reduction clause for the OpenMP parallelization. Reduction variables can be found by checking for critical regions in the compiler messages.
4. Find variables that need to be listed in the private clauses of OpenMP directives. An array must be private if the value of its index depends on any loop index in the loop nest. In addition, a variable must be private if it is declared in the loop body.
5. Insert appropriate OpenMP directives by adding necessary clauses such as reduction and private clauses into the serial code.

3 Performance Evaluation

3.1 Experimental Environments

In order to evaluate the effectiveness of the proposed OpenMP parallelization method, this paper uses three kernels of practical applications, called Antenna, BCM, and Earthquake kernels. These applications have been developed and/or optimized on NEC SX-9 for researches in the fields of science and engineering. Thus, the compiler messages generated by the NEC SX compiler are utilized for the proposed OpenMP parallelization method. The Antenna kernel is for the FDTD simulation of a lens antenna [4] and its Bytes/Flop ratio is 1.73. The BCM kernel is the main calculation of CFD with equally-spaced Cartesian meshes [5] and its Bytes/Flop is 8.0. The Earthquake kernel is for simulation of the seismic slow slip model [4] and its Bytes/Flop ratio is 2.0.

Four HPC systems whose specifications are shown in Table 1 are used for the performance evaluations. A parallelized OpenMP code is compiled by the compiler on each HPC system with OpenMP options as shown in Table 1. The original serial code is automatically parallelized by the compiler with enabling the highest-level optimizations except for the OpenMP support.

Table 1 Compile conditions of each HPC system

HPC system	Sockets	Cores	Compiler and options
NEC SX-9	16	1	sxf90 -Popenmp/-Pauto -Chopt -R2 (Rev.460)
Intel Nehalem EX	4	8	ifort -openmp/-parallel -xHost -O3 (Ver. 12.1.0 20110811)
Fujitsu FX10	1	16	frtpx -Kopenmp,noparallel/-Kparallel, noopenmp -Kfast (1.2.1)
Hitachi SR16000 M1	4	8	f90 -omp/-parallel -64 -model=M1 (Hitachi)

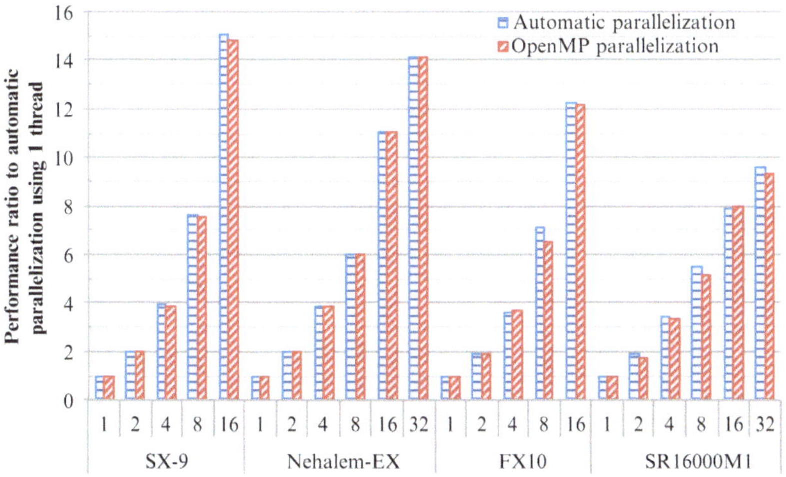

Fig. 1 Performance of the Antenna kernel

3.2 *Performance of OpenMP Codes Parallelized by using Automatic Parallelization Information*

Figure 1 shows the performance of the Antenna kernel code on each HPC system. The horizontal-axis indicates the HPC systems and the number of threads used in evaluation. The vertical-axis indicates the sustained performances that are normalized by the performance of the original code executed on each HPC system with a single thread. This figure shows that the performances of the OpenMP version are comparable to those of the original serial version on every HPC system. In the case of SX-9, since the proposed OpenMP parallelization method is performed by using the compiler messages of the NEC SX compiler, the same parallelization is applied to the OpenMP version. Since the sustained performance of the OpenMP version is the same as that of the automatically parallelized version, these results demonstrate that the proposed method enables a programmer to use OpenMP to express the parallelizations determined by the target compiler. In the cases of the other HPC systems, the Antenna code is also successfully parallelized by the automatic parallelization functions of their own compilers. Since the similar parallelization is applied for both the OpenMP version and the automatically parallelized version, the performances of these two versions are similar even when the number of threads increases. Although an overhead due to creating and deleting a team of threads can be observed, the overhead is sufficiently small.

Figure 2 shows the performance of the BCM kernel code on each HPC system. This figure shows that the performances of the automatically-parallelized version and the OpenMP version on SR16000M1 are similar. This is because automatic parallelization of the SR16000M1 compiler is similar to that of the NEC SX compiler. Furthermore, this figure clearly shows that the performance of the OpenMP version is obviously higher than those of the automatically parallelized versions in the cases

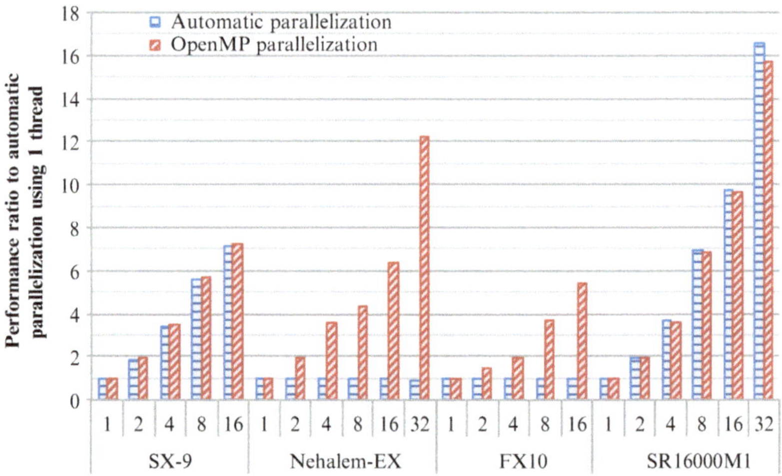

Fig. 2 Performance of the BCM kernel

of Nehalem-EX and FX10. The reason is that the compilers of Nehalem-EX and FX10 could not identify most of parallelizable loop nests. Since the dependencies among iterations cannot be analyzed, only 15.4 % of the parallelizable loops can be detected by their compilers. As the NEC SX compiler can analyze those loop nests and identify them as parallelizable loop nests, the proposed OpenMP parallelization method based on the compiler message can successfully parallelize such loop nests, and improve the performances of Nehalem-EX and FX10. These results clarify that the information from the target compiler is helpful for the OpenMP parallelization. Since other compilers may not be able to find those parallelizable loop nests, the OpenMP parallelization with the target compiler message is promising to improve the performance portability across different systems.

Figure 3 shows the performance of the Earthquake kernel code on each HPC system. Similar to the results of the other kernels, in the case of SR16000M1, as automatic parallelization by SR16000M1's compiler is similar to that by the NEC SX compiler even in the Earthquake kernel, its automatic parallelization and the proposed parallelization lead to similar sustained performance. In the case of Nehalem-EX, the performance of the automatically parallelized version cannot be improved even if the number of threads increases. This is because the compiler cannot automatically detect the parallelizable loop nests. The most dominant loop nest cannot be parallelized since the compiler decides that the parallelism of the loop nest is insufficient. Furthermore, in the cases of Nehalem-EX and FX10, even if a single thread is used for the execution, the performances of the automatically parallelized version are inferior to those of the OpenMP version. This is because automatic optimizations applied to the original version of the Earthquake kernel are different from those applied to the OpenMP version. Automatic SIMD optimizations by the compilers are applied only to the OpenMP version. By inserting the OpenMP

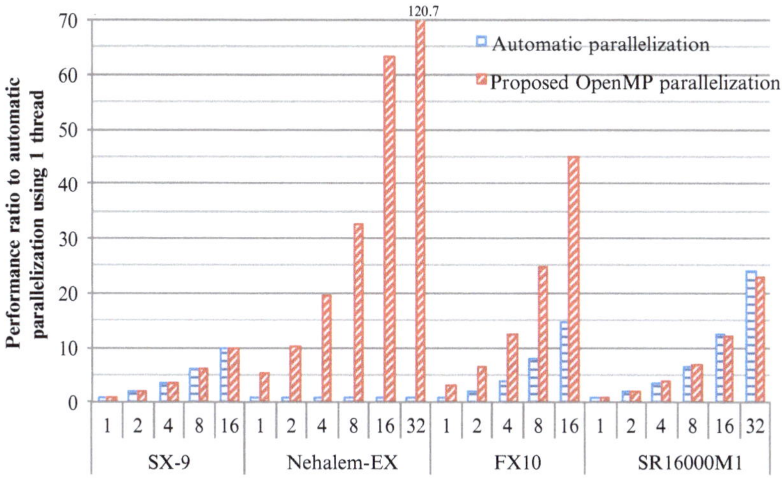

Fig. 3 Performance of Earthquake code

directives, as the search space of automatic optimizations for the proposed OpenMP version becomes narrower, further highly optimizations can successfully be applied only to the OpenMP code. Thus, the performances of the proposed OpenMP versions are superior to those of the automatically parallelized versions in Nehalem-EX and FX10.

Conclusions

As multiple and many core processors in recent HPC systems evolve, multi-thread parallelization becomes one of important factors to exploit the potential of the recent HPC systems. Use of automatic parallelization functions of a compiler is one of Major approaches to multithread parallelization. Due to the easiness of automatic parallelization, an HPC code is naturally and/or intentionally modified for the automatic optimization of a target compiler. Thus, the target compiler can effectively understand the structures and dependencies of the code, and apply automatic parallelization and optimizations. However, the performance of the code optimized for a target compiler is not always portable to different environments.

In this paper, by parallelizing a serial code with OpenMP, high performance portability of the code is realized. By using the compiler messages of automatic parallelization functions, the low-cost OpenMP parallelization method is proposed. By using compiler messages, the OpenMP parallelization can be easily applied to loop nests where are not automatically parallelized.

(continued)

Experimental results clarify that the compiler messages from the target compiler are useful for the OpenMP parallelization. Even in the case where other compilers are not able to find parallelizable loop nests, the proposed OpenMP parallelization method can appropriately parallelize the serial code by using the compiler messages. As a result, the performance of the OpenMP version is comparable to or higher than that of the automatically-parallelized version.

Acknowledgements This research was partially supported by Grant-in- Aid for Scientific Research (S) #21226018 and Core Research of Evolutional Science and Technology of Japan Science and Technology Agency (JST CREST) "An Evolutionary Approach to Construction of a Software Development Environment for Massively-Parallel Heterogeneous Systems".

References

1. Top 500 supercomputers sites. http://www.top500.org/
2. Mustafa, D., Aurangzeb, A., Eigenmann, R.: Performance analysis and tuning of automatically parallelized openmp applications. In: Proceedings of the 7th International Conference on OpenMP in the Petascale era, IWOMP'11, pp. 151–164. Springer, Berlin, Heidelberg. http://dl.acm.org/citation.cfm?id=2023025.2023041 (2011)
3. Himeno benchmark. http://accc.riken.jp/2444.htm
4. Soga, T., Musa, A., Shimomura, Y., Egawa, R., Itakura, K., Takizawa, H., Okabe, K., Kobayashi, H.: Performance evaluation of nec sx-9 using real science and engineering applications. In: Proceedings of the Conference on High Performance Computing Networking, Storage and Analysis, pp. 1–12 (2009). doi:10.1145/1654059.1654088
5. Nakahashi, K.: High-density mesh flow computations with pre-/post-data compressions. AIAA Paper **2005-4876** (2005)

EXTOLL and Data Movements in Heterogeneous Computing Environments

Holger Fröning

Abstract Driven by the end of Dennard scaling, we are observing a fundamental transition to communication-centric computing systems, based on heterogeneous processing units like CPUs, GPUs and others. The vast increase of parallelism and heterogeneity together with the advent of Big Data leads to systems in which communication is more expensive than computation, both in terms of energy and time. This work reviews EXTOLL as a new contender in the HPC interconnect arena, highly optimized for such environments. EXTOLL has been designed from scratch for HPC needs, allowing for slim and efficient software stacks. In the second part we will focus on GGAS, a research project we use to explore data movement optimizations in heterogeneous environments using EXTOLL.

1 Introduction

In the post-Dennard performance scaling, total chip power no longer remains constant, as threshold voltage scaling (and so supply voltage scaling) of transistors cannot be maintained. As a result, total chip power is increasing while it remained constant in the Dennard scaling era. Or in other terms, we can put more transistors on a die than we can afford to turn on, which leads to a huge need for execution models that optimize for energy consumption instead of time. As a result, in post-Dennard performance scaling the vast amount of performance increase originates to an increased concurrency, exposed by multi-/many-core processor and clustered architectures.

Power consumption is the main limitation of computing systems, no matter if they are small mobile devices that gear to maximize battery life, or if they are highly scalable clusters or datacenters. For the latter, power consumption is not only a technical limitation, also energy is expensive and already today the operational costs are dominated by energy and over the lifetime of such a system they can already exceed the initial acquisition costs.

H. Fröning (✉)
Institute of Computer Engineering, Ruprecht-Karls University of Heidelberg, Heidelberg, Germany
e-mail: holger.froening@ziti.uni-heidelberg.de; froening@uni-hd.de

M.M. Resch et al. (eds.), *Sustained Simulation Performance 2014*,
DOI 10.1007/978-3-319-10626-7_11

We can observe that technology diversity is pervasive. We can choose among a large variety of processors (CPU, GPU, MIC), storage (DRAM, FLASH, HDD, SSD), interconnects (10GE, 40GE, 100GE, Infiniband, EXTOLL ...) or even complete system architectures (IBM Blue Gene, Cray ...). Composing heterogeneous systems is mandatory due to power and performance reasons. In HPC, the performance and energy-efficiency of specialized processing units like accelerators has led to a wide use. A quick look at the Green500 list, which lists the most energy efficient computing systems in terms of their sustained GFLOPS/Watt metric, reveals that the top 10 entries are all based on heterogeneous architectures. While heterogeneous processing is pervasive, communication models and stacks are still solely designed and optimized for CPUs. Taking into account that in the future energy consumption will be dominated by data movements, heterogeneous and specialized communication models are mandatory to optimize overall energy efficiency.

As a result and following the transition to multi-/many-core, the computing landscape is undergoing another fundamental transition to communication-centric heterogeneous systems. Besides specialized processing, specialized communication paths are inevitable to maximize energy efficiency. In this work, two research efforts are presented and how they address the needs of this new era:

1. EXTOLL as a new contender in the area of scalable high performance interconnection networks, optimized for HPC and multi-core environments.
2. GGAS as a new communication model for an improved support of GPU-to-GPU data movements in terms of energy and time, which is based on a specialization of EXTOLL.

The remainder of this work is structured as follows. In the first part, EXTOLL and its features are presented. This is followed by an introduction to GGAS and how it builds on top of a specialized communication engine of EXTOLL. Related work is presented in section "Conclusion", while the last section concludes.

2 EXTOLL

EXTOLL started as a research project at the Computer Architecture Group at the University of Heidelberg, Germany, and is currently being productized by a spin-off company. Its overarching objective is to be highly optimized for the need of HPC. Besides the plain performance in terms of high bandwidth, low latency, special attention was put on an optimized and lean CPU-device interface, support for multi-core by efficient virtualization, a small memory footprint that allows for scalable communication, and finally an extremely high hardware message rate.

While most of the aspects above apply to the network interface, the network is based on direct topologies: 3D torus networks are naturally supported but are not mandatory. Any direct topology with a node degree of up to six is possible. EXTOLL supports lean communication stacks by optional in-order delivery of

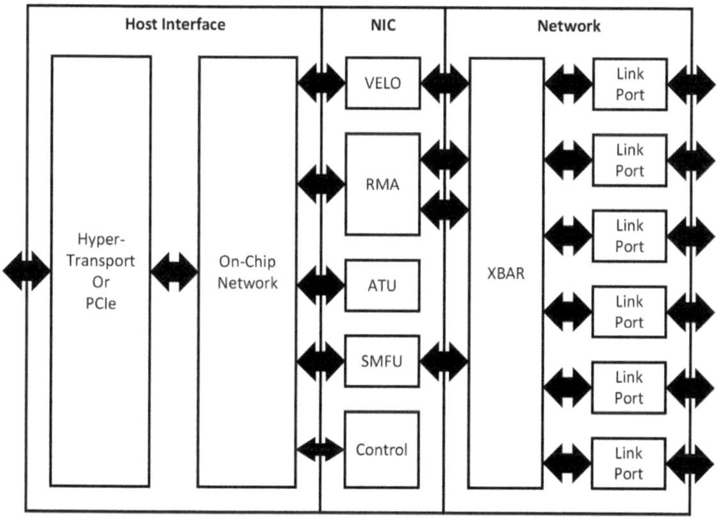

Fig. 1 EXTOLL top-level block diagram

Table 1 Different EXTOLL implementations and basic performance characteristics

	Ventoux	Galibier	Tourmalet
Technology	Xilinx Virtex-6 FPGA	Xilinx Virtex-6 FPGA	65 nm ASIC
Interface	Hypertransport	PCIe G1 x8	PCIe G3 x16
Node degree	6	4	6
Operating frequency (MHz)	200	156	750[a]
Internal data path width	64 bit	64 bit	128 bit

[a]Estimated

packets and a reliable transmission. The network switch is integrated into the add-in card, which provides six links towards the network. It employs virtual output queueing at switch level (VOQswitch) in order to improve head-of-line blocking. Routing is based on distributed tables and can be freely chosen to support arbitrary routing algorithms.

The following Fig. 1 shows the top-level block diagram of EXTOLL. It can be separated into three parts, the host interface that is either based on PCIe or HyperTransport, the network interface that includes different communication engines, and the network part that mainly consists of the router and link controllers.

Note that the EXTOLL is highly modular and configurable, thus various incarnations already exist and likely more will come in the future. This can apply to different aspects, up to now we have seen different technologies, host interfaces, node degrees, and communication engines. Different topologies also come with different characteristics like data path width and operating frequency; the following Table 1 summarizes the most important incarnations yet.

2.1 Communication Engines

EXTOLL is based on different communication engines for different needs. Speaking about HPC, the most important interface to applications is the Message Passing Interface (MPI) that relies on a message exchange for communication and synchronization purposes. As we will see later, certain use models might benefit from more specialized engines, which can be easily integrated into the EXTOLL framework.

Communication engines are usually fully virtualized to efficiently support a virtually unlimited number of end points per node (i.e., up to 100 s of processes or threads). Also, support for user-level communication to bypass costly OS interactions or hypervisors is usually included. Security and safety is guaranteed by appropriate features in the communication engines that prevent work requests to disable networking components.

Here, we will shortly review the two most important communication engines for message passing: short messages are usually handled by the VELO engine, while bulk transfers naturally map to the RMA engine.

2.1.1 Support for Small Messages by VELO

VELO's key idea is to send and receive network messages with as little overhead in terms of time as possible, making it in particular suitable for short data transfers. Essentially, one Programmed IO (PIO) write to VELO is sufficient to send out a message. All necessary information including header and payload is included in this write operation. The write combining buffer of the CPU usually ensures that multiple store instructions are coalesced into a single write operation, if this fails due to the scheduling effects or similar the VELO unit ensures that multiple transactions are aggregated before forwarded.

Accesses to main memory are costly and can be completely avoided this way. Successful work request generation can be controlled directly after the PIO write or later by employing a lazy update scheme.

On the receive side, VELO leverages buffers in main memory to temporarily store incoming packets. These buffers or mailboxes are organized as ring buffers with fixed-size slots. This static partitioning allows processes or threads to poll on predetermined memory locations when waiting for new packets. The coherence protocol ensures that the number of cache misses during this costly polling period is minimized:

1. The receiving process calculates the address for the next incoming packet within the ring buffer. The first load targeting that address will likely result in a cache miss.
2. Upon the next incoming packet, the VELO unit updates this slot of the ring buffer. The memory coherence protocol invalidates copies of these addresses in the CPU caches.

3. Subsequent loads of the process still target the same address, thus will result in cache hits.
4. The next load of the polling process results again in a cache miss, bringing the most recent value from memory to the computing core.

As can be seen, independent of the polling period in the best case only two cache misses are observed. Scheduling or process migration might increase this though, but for the common case memory traffic is minimized.

The only disadvantages of this approach is that message size is limited by the maximum size of a PIO transaction which is 64 B. This already includes an 8 B header. For sending larger amounts of data, the processor has to subsequently write portions using this PIO scheme, which limits overlap opportunities. For smaller payloads these two effects are not cumbering, so usually VELO is employed for payloads up to 1 kB.

2.1.2 One-Sided Bulk Transfers by RMA

The main objective of the RMA engine is to support Put/Get semantics with an advanced notification scheme. Notifications for such one-sided operations are mandatory to ensure consistency and joint synchronization points, and often implementations tend to not optimize this important aspect.

The communication path of RMA is threefold. The requester unit sends a work request, which could be either a Put or Get request, to the target node. For a Put request, payload is already included in the request and it is forwarded to the completer unit on the target side. For Get request, the responder unit on the target side is responsible to fetch data and to respond by sending back the data. For all three units, completion notifications can be generated, allowing upper software stacks to track the progress of each step.

The Put/Get semantics of RMA are based on a global address space by using Network Logical Addresses (NLAs). NLAS are generated by registering memory regions, and during this process the associated physical pages are also pinned in main memory. The RMA engine is assisted by an Address Translation Unit (ATU), which allows address translations by caching and page table walks autonomously on the network device. Thus, for Put/Get operations the OS can completely be bypassed. Only for the registration process system calls are required.

Similarly to VELO, the processor-network interface is highly optimized for low overhead and bus traffic. Work request issue is based on writing descriptors directly to the device. We opted out a use of intermediate queues in main memory as they require additional signalling of work requests to the device (doorbell functionality). The notification scheme has many similarities to the VELO mailbox scheme, and thus polling for completion notifications results in minimal traffic by exploiting caching possibilities, too.

2.1.3 MPI Design Issues

MPI is a complex messaging protocol and library that abstracts most communication details from the user. For send/receive semantics, the underlying transport mechanism is completely hidden from the user. Depending on the payload size, MPI libraries usually employ two transport mechanisms: eager for small-sized transfers and rendezvous for larger bulk transfers. For eager communication EXTOLL uses the VELO communication path, leveraging the low overhead (fast work request issue, no registration required).

Similarly, the initial rendezvous message of the rendezvous protocol is also handled by VELO. This protocol will only pull data from the source node when appropriate buffers are allocated on the target node, and for this bulk data transfer the EXTOLL Get operation is most suitable. The advanced notification scheme of RMA allows to signal completion to both sender and receiver with as little overhead as possible.

In addition, MPI libraries highly benefit from EXTOLL's properties like in-order delivery of packets and reliable transmission. No efforts have to be done regarding buffering for re-plays or re-sorting of incoming packets, significantly easing and slimming this software stack.

2.2 Key Performance Characteristics

EXTOLL's main objectives include high performance in terms of sustained bandwidth and latency. While high bandwidth is usually reached for rather larger payloads, and latency does not fully describe the throughput available for short data transfers, the message rate is a good metric to describe this case. The message rate essentially describes how many messages can be sent out per second, and usually this metric peaks out for rather small transfers in the range of 8–32 B and for concurrent streams by employing multiple process pairs on two nodes.

EXTOLL is an architecture with many different incarnations in terms of FPGA and ASIC technologies. FPGAs are reconfigurable devices but are inferior to ASICs in terms of capacity and operating frequency. Up to now, several FPGA technologies have been evaluated but no ASIC version yet. This is currently in production, but based on simulation runs we can estimate performance numbers. The following Table 2 summarizes key characteristics of different incarnations and their corresponding performance numbers.

2.3 Additional Reading

More information about EXTOLL can be found in various publications. For VELO please refer to [12] and for RMA to [19]. The message rate of an earlier version

Table 2 Different EXTOLL implementations and basic performance characteristics

	Ventoux	Galibier	Tourmalet[a]
Technology	Xilinx Virtex-6 FPGA	Xilinx Virtex-6 FPGA	65 nm ASIC
Half round-trip latency (μs)	<1.0	1	0.4–0.6
Hop latency (ns)	<150	<150	<60
Peak bandwidth (GB/s)	1.6	1.6	>10
Message rate (M/s)	>25	>25	>100
Memory registration latency (μs)	∼2	∼2	∼2

[a]Estimated performance numbers

of EXTOLL has been extensively studied in [8]. The book chapter [18] gives a comprehensive overview of the details of EXTOLL and its FPGA implementation. A dynamic software flow control scheme for the VELO unit has been proposed and evaluated in [25].

EXTOLL was also used in the MEMSCALE project as a shared memory mapper for transparent address space extensions. More information about the concept can be found in [13–15], more details in particular on in-memory databases in [16, 17], and an overview in [9].

The use of EXTOLL for data-warehousing workloads on distributed heterogeneous systems is described in [28]. This work focuses on a runtime for such systems to assist the user in registration and one-sided data movement operations.

3 Global GPU Address Spaces

GPU Computing is pervasively used in High Performance Computing to reduce execution times and to improve energy efficiency. In this section we will see how GPU architectures differ from CPUs, and which problems arise when they are employed in clusters. To overcome these limitations we are proposing Global GPU Address Spaces (GGAS) as a communication model that is in-line with the GPU's execution model.

3.1 GPUs and Accelerated Clusters

GPUs are powerful high-core-count devices that are based on multiple Streaming Multiprocessors (SMs), with each of them executing thousands of threads concurrently. A SM is composed of a large number of computing cores, load/store units and a shared scratchpad memory. A thread block can execute within the domain of an SM. The scheduling leverages the parallel slackness of the vast amount of threads to hide long-latency events like memory accesses. Threads are scheduled in so-called warps of (currently) 32 threads. Threads within a warp share the same instruction stream, thus divergent branching leads to performance penalties as the results of inactive branches have to be nullified.

While GPUs have been primarily designed for graphics processing, today the play an important role as accelerators. This trend has been boosted by the introduction of domain-specific languages like CUDA or OpenCL, which allow programming GPUs without any background in graphics processing. In these programming models, parallel computational kernels are executed on the GPU, while the host CPU is responsible for launching kernels and controlling data movements from main memory to accelerator memory and vice versa.

A major difference between the execution model of general-purpose CPUs and specialized GPUs is the degree of concurrency. CPUs prefer few heavy-weight threads that execute autonomously and are rather costly to start and switch. For these reasons, usually the same amount of threads as the order of magnitude of physical computing cores are started. GPUs however rely on the parallel slackness of having many more threads executing than physical computing cores are present. GPUs can switch threads at negligible costs, allowing to hide long-latency events. A common approach is to maximize the number of concurrently executing threads to maximize scheduling opportunities, so usually for each output element a dedicated thread is started.

Similarly to the massive concurrency for computations, memory accesses are also based on many threads working collaboratively. Data movements from device memory to the scratchpad and vice versa are usually performed by many threads in a collaborative way, having each thread only move one element. As a result, the execution model is quite simple as all tasks, including data movement and computing, have to be partitioned as fine-grained as possible.

However, for GPU clusters this simplicity is lost. Communication stacks are designed for CPUs and a single-threaded execution model, making them incompatible with GPUs. In particular, GPUs perform poorly when constructing work requests to be issued to the network device. Polling on completion notifications is performing similarly bad. In general, the message-passing communication model is not in-line with the GPU's execution model, resulting in hybrid programming models like MPI+CUDA, MPI+OpenCL or similar. Complexity is amplified and the beauty of simplicity of the GPU's execution model is lost.

As GPUs are matching poorly the existing communication models, developers rather opt to revert the control flow back to the CPU domain for messaging tasks, and restart GPU execution by new kernel launches. Such context switches are costly in terms of time, and in addition the CPU cannot be left idle which results in additional energy consumption.

3.2 A Thread-Collaborative Communication Model

A possible solution to overcome this mismatch between the GPU execution model and existing communication models is a transparent extension of the local address

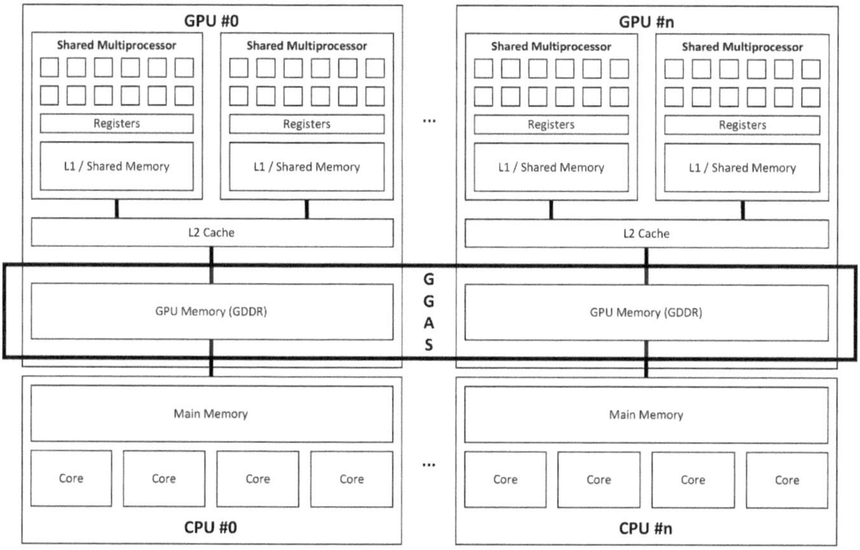

Fig. 2 A GPU global address space spanned across distributed GPU device memory

space. This extension maps remote memory resources into the local address space, allowing the local GPU to access remote resources like device memory with native load/store instructions. From a GPU's perspective, there is no difference between a local main memory access and an access to remote GPU memory, and this transparency allows to maintain the commodity aspect of GPUs and the associated cost-effectiveness.

Figure 2 depicts how such a global address space is used to aggregate distributed GPU memory. Any GPU in the system can use it to access all memory resources, independently if local or remote. This allows for sharing, communication and synchronization. The accesses should be performed in the same way as local device memory is accessed, thus in a thread-collaborative way. Note that locality is not necessarily exposed to the user, but abstractions should allow for appropriate locality optimizations in order to address performance disparities in terms of latency and bandwidth.

A transparent global address space allows for various communication models, ranging from compiler-driven automated data partitioning and communication to library-based messaging abstractions. Our initial work is based on the latter, and we set up a mailbox scheme with send/receive semantics to communicate between distributed GPUs [10]. However, mailbox update and synchronization is still based on many threads performing these tasks collaboratively, for instance one thread within such a collaborative thread array might be responsible for a single element only.

Table 3 Basic GGAS performance characteristics

	GGAS	MPI
Peak bandwidth (MB/s)	800	755
Bandwidth @ 1 kB (MB/s)	780	70
Half round-trip latency (μs)	1.9	13.8

3.3 *Key Performance Characteristics*

In the following Table 3 we summarize the key performance characteristics from our experiments. We chose to compare against MPI as state-of-the-art solution, but we include additional copies from GPU memory to staging buffers in host memory (based on cudaMemcpy operations) for a fair comparison.

Note that absolute performance levels are limited by the employed FPGA technology. Experiments have been conducted using Galibier cards, which are summarized regarding limitations in Table 2. We are looking forward to re-evaluate GGAS using the coming ASIC incarnation of EXTOLL, as well as a to explore a wider set of workloads.

Energy consumption is another interesting aspect, as GGAS allows for the first time to bypass CPUs for communication tasks, without sacrificing performance. Experiments have shown that the idling CPUs and the reduced run time of GGAS allows reducing overall energy consumption when comparing against MPI as state-of-the-art. For a global allreduce operation, energy consumption is reduced by 47 %; for a global sum it is reduced by 44 % [22].

3.4 *Additional Reading*

The GGAS concept has been proposed in [20], and initial results are compared against an Infiniband network. The underlying communication engine to forward load/store operations is described in [7]. We explored the impact of GGAS on collective operations in detail in [22], in particular the aspect of energy efficiency. A similar work but focussing on controlling Put/Get abstractions from a GPU device is presented in [21].

4 Related Work

Various related work in the area of interconnection network is present, a comprehensive listing here is prohibitive due to space limitations. Currently the most important interconnect for HPC is likely the Infiniband standard, in particular the ConnectX technology by Mellanox [27]. While EXTOLL is designed from scratch for HPC workloads, Infiniband was designed as a general-purpose network and is therefore less specialized.

Other important networks for HPC include IBM Blue Gene [11], IBM PERCS [4], Cray Seastar [5] Cray Gemini [2] and Cray Aries [6]. All of them excel in terms of performance and scalability, however they are not sold separately but instead are only available with the corresponding computing systems. One of EXTOLL's goal is to close the gap between such specialized but non-commodity systems and standard clusters.

Regarding GPU-to-GPU communication efforts, most related work relies on CPU assistance for communication tasks. Examples include [3, 23, 24, 29]. A portable programming framework for GPU clusters is introduced in [1], but again it relies on the CPU to handle communication tasks. DGCN [26] is an approach that offers send/receive semantics for GPUs, but these abstractions again trigger the host CPU. Note that the authors of this work demand for a direct communication technique like it is made possible by GGAS.

Conclusion

Heterogeneous processing has become pervasive for reasons that include increased performance and improved energy efficiency, as the processing units not only differ in their raw computational power, but also in their energy efficiency due to specialization. As data movements will increasingly contribute to the overall energy consumption, it is mandatory to optimize such operations in terms of energy efficiency too. As an example for such a specialized communication model we are proposing GGAS, which is highly in-line with the GPU's execution model, making it a natural fit for such thread-collaborative processors. EXTOLL's flexible architecture has allowed an easy integration of the specialized communication engine for GGAS, making it a prime example for specialized communication models and their impact on data movement efficiency in terms of energy and time.

Future work will explore more data movement optimizations, with an increasing focus on energy efficiency. Up to now, experiments were limited by the lack of a scalable computing systems. We are currently setting up the `High Octane` system, which allows for customizable communication paths and models and will consist of 8–16 interconnected GPUs, each paired with a Xeon-type CPU and an EXTOLL network device. This provisioning makes this system very communication-centric, and its scale and various communication paths allow for various experiments and explorations.[1]

[1] Another goal of High Octane is to be an open platform for experiments, explorations and developments in the area of communication-centric heterogeneous systems. If you are interested in exploring your application on a heterogeneous system with regard to different communication models and paths, we recommend to contact us.

Acknowledgements The author would especially like to thank Sudhakar Yalamanchili for extensive discussions about the future of computing in terms of trends, opportunities and challenges. Lena Oden and Benjamin Klenk have significantly contributed to the success of the GGAS research effort and provided valuable feedback and insights. EXTOLL is a joint research project that has been driven by various people, including Ulrich Brüning, Mondrian Nüssle and many more past and present members of the Computer Architecture Group of the Institute of Computer Engineering at the Ruprecht-Karls University of Heidelberg. Last, the sponsoring of these two projects by Nvidia and Xilinx is highly acknowledged.

References

1. Aji, A.M., et al.: MPI-ACC: An integrated and extensible approach to data movement in accelerator based systems. In: International Conference on High Performance Computing and Communications (HPCC 2012) (2012)
2. Alverson, R., Roweth, D., Kaplan, L.: The Gemini system interconnect. In: 18th IEEE Symposium on High Performance Interconnects (HOTI 2010) (2010)
3. Ammendola, R., et al.: GPU Peer-to-Peer techniques applied to a cluster interconnect. In: International Symposium on Parallel and Distributed Processing Workshops (IPDPSW 2013) (2013)
4. Arimilli, B., et al.: The PERCS high-performance interconnect. In: 18th IEEE Symposium on High Performance Interconnects (HOTI 2010) (2010)
5. Brightwell, R., Pedretti, K., Underwood, K.D.: Initial performance evaluation of the Cray SeaStar interconnect. In: 13th Symposium on High Performance Interconnects (HOTI 2005) (2005)
6. Faanes, G., et al.: Cray cascade: a scalable HPC system based on a Dragonfly network. In: International IEEE Conference on High Performance Computing, Networking, Storage and Analysis (SC 2012) (2012)
7. Fröning, H., Litz, H.: Efficient hardware support for the partitioned global address space. In: 10th Workshop on Communication Architecture for Clusters (CAC2010), in conjunction with 24th International Parallel and Distributed Processing Symposium (IPDPSW 2010) (2010)
8. Fröning H., Nüssle, M., Litz, H., Leber, C., Brüning, U.: On achieving high message rates. In: 13th IEEE/ACM International Symposium on Cluster, Cloud and Grid Computing (CCGrid 2013) (2013)
9. Fröning, H., Silla, F., Montaner, H.: MEMSCALE: Re-architecting memory resources for clusters. In: Vanderbauwhede, W., Khaled, B. (eds.) High Performance Computing Using FPGAs. Springer, New York (2013)
10. Klenk, B.: Comparing different communication paradigms for data-parallel processors. M.Sc. thesis, Institute of Computer Engineering, Ruprecht-Karls University of Heidelberg, Heidelberg (2013)
11. Liang, Y., et al.: BlueGene/L failure analysis and prediction models. In: International IEEE Conference on Dependable Systems and Networks (DSN 2006) (2006)
12. Litz, H., Fröning, H., Nüssle, M., Brüning, U.: VELO: A novel communication engine for ultra-low latency message transfers. In: 37th International Conference on Parallel Processing (ICPP 2008) (2008)
13. Montaner, H., Silla, F., Fröning, H., Duato, J.: Getting rid of coherency overhead for memory-hungry applications. In: IEEE International Conference on Cluster Computing (CLUSTER 2010) (2010)
14. Montaner, H., Silla, F., Fröning, H., Duato, J.: A New Degree of Freedom for Memory Allocation in Clusters. Cluster Computing. Springer, Hidelberg (2011)

15. Montaner, H., Silla, F., Fröning, H., Duato, J.: Unleash your memory-constrained applications: a 32-node non-coherent distributed-memory prototype cluster. In: 13th IEEE International Conference on High Performance Computing and Communications (HPCC 2011) (2011)
16. Montaner, H., Silla, F., Fröning, H., Duato, J.: MEMSCALE: a scalable environment for databases. In: 13th IEEE International Conference on High Performance Computing and Communications (HPCC 2011) (2011)
17. Montaner, H., Silla, F., Fröning, H., Duato, J.: MEMSCALE: in-cluster-memory databases. In: 20th ACM Conference on Information and Knowledge Management (CIKM 2011) (2011)
18. Nüssle, M., Fröning, H., Kapferer, S., Brüning, U.: Accelerate communication, not computation! In: Vanderbauwhede, W., Khaled, B. (eds.) High Performance Computing Using FPGAs. Springer, New York (2013)
19. Nüssle, M., Scherer, M., Brüning, U.: A resource optimized remote-memory-access architecture for low-latency communication. In: 38th International Conference on Parallel Processing (ICPP 2009) (2009)
20. Oden, L., Fröning, H.: GGAS: Global GPU address spaces for efficient communication in heterogeneous clusters. In: IEEE International Conference on Cluster Computing (CLUSTER 2013) (2013)
21. Oden, L., Fröning, H., Pfreundt, F-J.: Infiniband-verbs on GPU: a case study of controlling an Infiniband network device from the GPU. In: 4th International Workshop on Accelerators and Hybrid Exascale Systems (AsHES), in Conjunction with 28th International Parallel and Distributed Processing Symposium (IPDPSW 2014) (2014)
22. Oden, L., Klenk, B., Fröning, H.: Energy-Efficient Collective Reduce and Allreduce Operations on Distributed GPUs. In: 14th IEEE/ACM International Symposium on Cluster, Cloud and Grid Computing (CCGrid 2014) (2014)
23. Panda, D.K.: MVAPICH2: A high performance MPI library for NVIDIA GPU clusters with InfiniBand. In: GPU Technology Conference (GTC 2013) (2013)
24. Potluri, S., et al.: Optimizing MPI communication on multi-GPU systems using CUDA interprocess communication. In: International Parallel and Distributed Processing Symposium Workshops (IPDPSW 2012) (2012)
25. Prades, J., Silla, F., Duato, J., Fröning, H., Nüssle, M.: A new end-to-end flow-control mechanism for high performance computing clusters. In: IEEE International Conference on Cluster Computing (CLUSTER 2012) (2012)
26. Stuart, J.A., Owens, J.D.: Message passing on data-parallel architectures. In: IEEE International Symposium on Parallel and Distributed Processing (IPDPS 2009) (2009)
27. Sur, S., Koop, M.J., Chai, L., Panda, D.K.: Performance analysis and evaluation of Mellanox ConnectX InfiniBand architecture with multi-core platforms. In: 15th IEEE Symposium on High Performance Interconnects (HOTI 2007) (2007)
28. Young, J., Merritt, A., Shon, S.H., Yalamanchili, S., Schwan, S., Fröning, H.: Oncilla: a GAS runtime for efficient resource allocation and data movement in accelerated clusters. In: IEEE International Conference on Cluster Computing (CLUSTER 2013) (2013)
29. Wang, H., et al.: MVAPICH2-GPU: Optimized GPU to GPU communication for InfiniBand clusters. In: International Supercomputing Conference (ISC 2011) (2011)

Requirements for Modern Network Infrastructures

Jens Aßmann, Alexander Kiontke, and Sabine Roller

Abstract Modern networks are currently undergoing significant changes and have to deal with rapidly changing requirements. Especially with an alternate working behavior of users it is necessary to adjust the network services and management. This requires more sophisticated traffic engineering. This paper reviews the current state of the art and shows the application of new strategies to a mid-size German university. Practical needs and theoretical concepts are brought together to develop a modern network management infrastructure.

1 Motivation

Modern networks are currently undergoing significant changes and have to deal with rapidly changing requirements. Especially with an alternate working behavior of users it is necessary to adjust the network services and management. While in the past users basically worked with an end device placed at the desk, nowadays it is usual to work with variable devices at different places with varying capabilities to get access to the network (e.g. LAN, WLAN, mobile access). However, the users require an equal network behavior, thus it is necessary to provide a consistent feature set independent of device type, location and kind of access. In particular in relation to network security and manageability, Bring Your Own Device (BYOD) is a big challenge for network administrators.

Another requirement is to reduce prime and maintenance costs. Thus, an optimal utilization of network resources (e.g. bandwidth, ...) is essential. As a result the network has to provide segregated services on a common infrastructure which makes the use of virtualization and overlay technologies mandatory. Certainly this leads to a growing complexity with which university networks have to deal.

The University of Siegen is a mid-size public university in Germany. In the year 2014, approx. 20,000 students and 2,455 employees are working at the university. The university is distributed over eight locations all-over the city. Until the end of

J. Aßmann (✉) • A. Kiontke • S. Roller
University of Siegen, Center for Information and Media Technology, Siegen, Germany
e-mail: jens.assmann@uni-siegen.de; alexander.kiontke@uni-siegen.de; sabine.roller@uni-siegen.de

M.M. Resch et al. (eds.), *Sustained Simulation Performance 2014*,
DOI 10.1007/978-3-319-10626-7_12

2014, two more locations will be constituted. In the past, the distributed locations were connected via directed radio in a ring topology. By the end of 2014, it will be replaced by optical fibre. The locations are tenanted by different institutes, departments, and faculties, from humanities to engineering, from arts to high energy physics. Therefore, they have disparate and varying demands on the network infrastructure and services. This leads to an increasing heterogeneity. Suitable traffic engineering (TE) is necessary to get the most benefit and optimal efficiency from the available resources.

This section outlines related work including the advantages and the applicability in the university.

1.1 MPLS Traffic Engineering in OSPF Networks a Combined Approach

The research report of Köhler and Binzenhöfer [1], shows that a combination of multi-protocol label switching (MPLS, see [2]) and an interior gateway protocol (IGP) like open shortest path first (OSPF) [3] allows to achieve a fast linear optimization.

The publication presents a concept named DDR:

- Decompose the network.
- Design the MPLS overlay network.
- Re-assemble the network.

The advantages of a combined approach are shown in this paper and the observations that have been made are promising. The focus of this research report is TE. To apply it to the settings at the University of Siegen, an MPLS infrastructure is to be built, not only for TE, but also to be able to offer centralised services.

1.2 Enabling Software Defined Network (SDN) in Old School Networks with Software-Controlled Routing Protocols

The authors of [4] propose to inject OSPF messages into the network to adjust the routing decisions. With this technique they want to add external information into the network to optimize the traffic flow. This work tries to implement a software defined network (SDN) via standardized and well understood OSPF messages.

2 Requirements for Modern Network Development at the University

This section describes the requirement for modern network development. The University of Siegen is exemplary for progress and future development of university networks. New structures have to be implemented in existing infrastructures as well as management tools. Current structures therefore determine the target settings. Thus, the situation at the university in the past and the actual planning base for the years 2014–2016 are described to derive exemplified requirements.

The following subsections start with a description of the architecture, afterwards the requirements and some decision criteria are shown. The subsections continue, if applicable, with the security paradigm and end with an evaluation of benefits and downsides.

2.1 Collision Domain

We started with a flat two-tier architecture. As shown in Fig. 1, there was one core router, which was responsible for VLAN (virtual local area network) management, firewalling, and routing. The second tier was comprised of the access switches. These switches were connected in a ring topology which was one reason to use spanning tree for loop detection. The (rapid) spanning tree protocol creates a loop-free spanning tree inside a layer 2 domain by defining multiple role states [5].

As already mentioned, the university is distributed over different locations. Due to office space limitations, the members of a single lab are sometimes located at different sites. Thus, the requirement to connect single offices that belong to one lab at different locations had to be implemented in the past into the same layer 2 domain. (We separate different labs in layer 2 domains.) The management in this architecture was simple, the reliability was very poor. On the other hand, this structure allowed us to deploy every VLAN on the whole campus without the need to change anything in the core network.

The different locations were connected by directed radio. With respect to management requirements, connections via directed radio links have some disadvantages: Latency, jitter, drops and no reliable Quality of Service (QoS) mechanism. They are also very weather dependent regarding the round trip time, the reliability and the bandwidth.

The firewalling was centralized on one non-stateful firewall. While this provides good maintainability, this setup was also a single point of failure.

The reliability was mainly impacted by the spanning tree algorithm (loop prevention) and the variable bandwidth. There was no way to define a maximum data transfer rate on this medium without wasting bandwidth to a large scale. Without such a bandwidth definition it is not possible to define drop thresholds (The switch does not know when it should start dropping dispensable packets.).

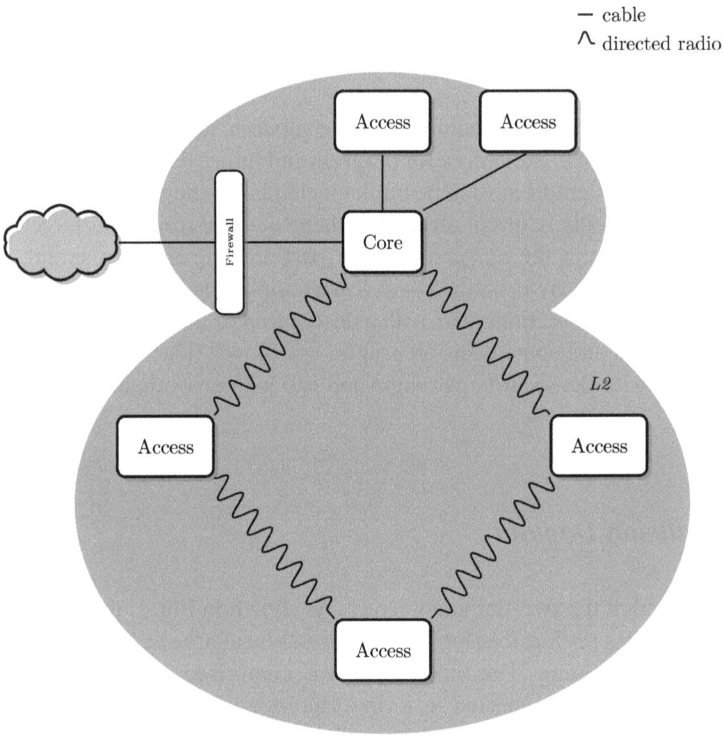

Fig. 1 Two-tier architecture with directed radio links and a single ISP connection

The unreliable connection in the ring architecture combined with the high re-convergence time of the spanning tree protocol was a reason for disconnections throughout the whole university several times. In a bad case scenario this can disable the whole network infrastructure for more than 90 s.

2.2 Routing in the Core

To increase reliability, we changed the two-tier architecture to a three-tier architecture which is more reliable in our setup. As shown in the Fig. 2 there is now a distinguished router for every location. This so called distribution layer is implemented between the core and the access layer. The connection in the core network is now a routed connection using the OSPF protocol. We also use first hop router redundancy protocols (FHRP) like hot standby router protocol (HSRP) and gateway load balancing protocol (GLBP), both Cisco proprietary. Which protocol is used depends on the local network requirements. This enables us to use a redundant router setup with more than one router as active forwarder (even in HSRP).

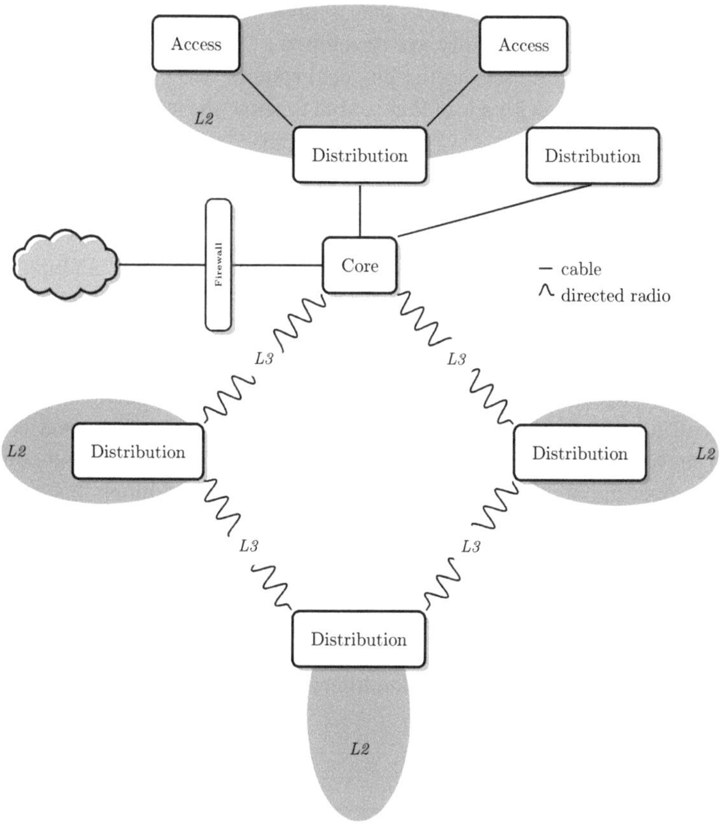

Fig. 2 Routed three-tier architecture with directed radio links and a single ISP connection

This routed connection is able to balance the load through the ring via the OSPF distances. This is the first TE approach to distribute the load equally.

There is now a separate layer 2 area for every location. These layer 2 domains are smaller and there are no unreliable connections between the switches.

For security purposes, there is a centralized stateful firewall at the provider edge (see Sect. 2.1). This is also shown in Fig. 2. The internal firewalling is done on every router with a non-stateful packet filter mechanism.

The management in this setup gets complex if a department is distributed over more than one single location or when there are different access methods like LAN, WLAN or mobile access. When this occurs, the users are in different IP subnets. For this reason, there might be multiple rules in the different firewall systems with different source IPs belonging to a single access policy. Services that rely on layer 2 connectivity do not work in this setup.

The time and effort required to find misconfiguration in this network is increasing. We built several tools to deal with this complexity. We created dynamic maps, that show the actual OSPF distances and the traffic; this allows us to manually re-adjust the costs. This is a process that is time consuming and not scalable. To achieve a better debugging ability, we developed a "traceroute clone", which maps the different routing addresses to the physical routers. We document core changes also in a graphical way with a tool that is able to show the state of the core routing.

2.3 Routing with Redundant ISP Connection

The approach described in Sect. 2.2 had one single point of failure, because the connection to the ISP was not redundant. This was changed to increase the reliability. We used the proven three-tier architecture and added another firewall to the provider edge (Fig. 3). We do not use these connections *active/backup*, but *active/active*. This allows us to use the double bandwidth with a higher redundancy level. With a redundant *active/active* ISP connection, there is a good chance to route in an asymmetric way. Therefore traffic may leave through one connection and return over the other one. It can not be in our interest to block this asymmetric routing, because we would need very fine granular routing entries in the exterior and in the interior routing protocol. This would be error-prone, time consuming and result in a large routing table.

Another requirement is a transparent centralized web filter technology to counteract attack scenarios against end user computer (e.g. drive by downloads). To use a transparent web filter the traffic to the filtered network must pass the web filter in

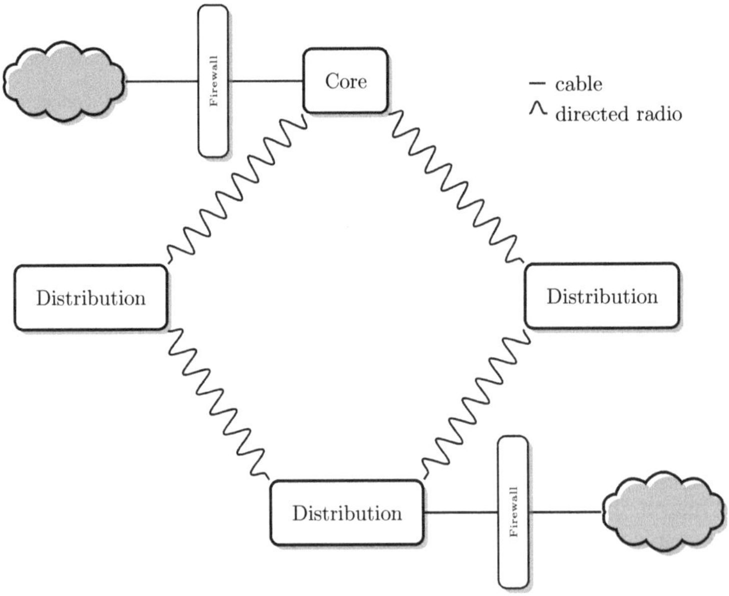

Fig. 3 Routing with redundant ISP connection

the outgoing and in the incoming direction. In an asymmetric routing case this can only be ensured when these web filters are attached to the distribution layer from a single location. Currently, we are using a redirection technology which requires layer 2 adjacency.

In an asymmetric case, some other security mechanisms like stateful firewalling (on the provider edge) also need special treatment. To create a state in a firewall, this firewall must see the initial connection handshake. In an asymmetric case the firewall will only see one direction of traffic. There are some techniques to synchronize firewall states but these have high requirements on bandwidth and reliability. Therefore, these are not applicable in settings with directed radio links.

With the second ISP connection we achieved the double bandwidth and a higher availability, but we also added another layer of complexity. We now have two different routing protocols, that are synchronized. Stateful firewalling on the provider edge is not possible at the moment. As a workaround, we are using packet filters at these points. Stateful firewalling is implemented at each location. This method does not scale, we have, due to the already mentioned different requirements in the locations, unequal bandwidth-requirements. Firewalls are an expensive resource, thus they should not lie idle. Therefore a centralized firewall would be a better approach, but it is not applicable in this architecture.

To lower the complexity, we developed maps that are able to show the current announced routing table. This enables us to determine the future packet flow.

2.4 Optical Fibre

The next step is to deploy fibre connections and replace the directed radio stations. As shown in Fig. 4, the fibre infrastructure will also be deployed in a ring. With this infrastructure we want to achieve reliable and fast connections throughout all locations. We plan to use dense wavelength division multiplexing (DWDM) to multiply the "physical" connections. This technique adds another abstraction layer, that needs to be managed.

These conditions enable us to use a stateful firewall at the provider edge.

This will not solve the firewall scaling problem at the single location. To reproduce and find misconfiguration on the network, there is the need for a management solution that should be able to collect the actual network state, provide some kind of history and should be expandable.

2.5 Optical Fibre with MPLS

To increase firewall scalability, we plan to use a central firewall design.

In an architectural deployment like Fig. 5 there is no need to worry about asymmetric routing. We want to use a tunnel technology like MPLS/VPN to route

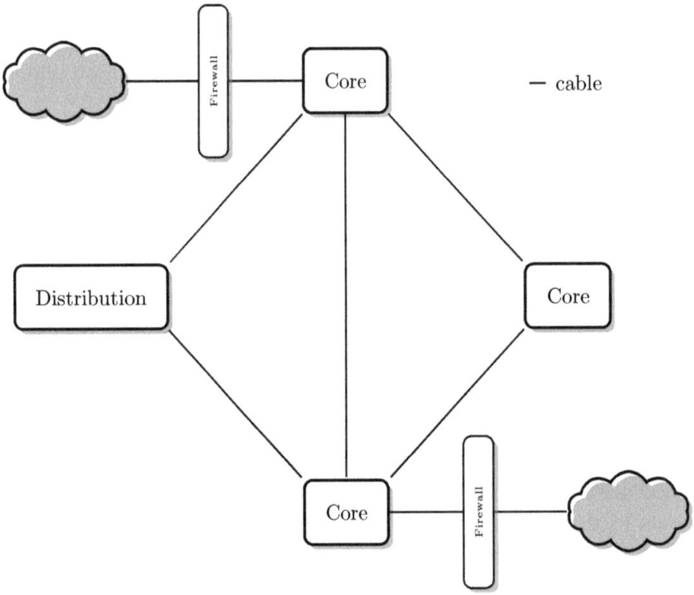

Fig. 4 Designed fibre ring with redundant ISP connection

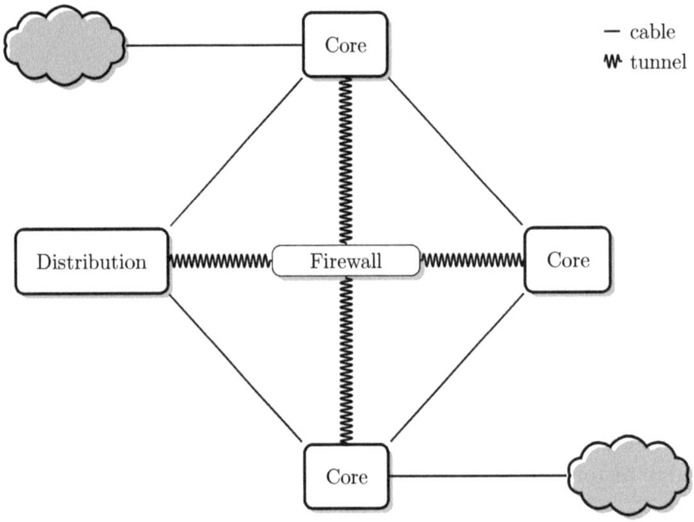

Fig. 5 Designed fibre ring with centralised firewall

traffic through the central firewall system. This system may be implemented as a firewall cluster. Such a cluster is able to allocate resources dynamically.

The combination of OSPF and MPLS was already shown in 2003 in [1]. This research report did not cover a more flexible TE solution that would be able to use external information for routing decisions (e.g. more SDN-like)

We mentioned in Sect. 2.3 that there are techniques to synchronize firewall states in a redundant setup, even when these are not at the same location. These techniques have certain disadvantages: At first, they need much bandwidth to achieve synchronization and they also need separated links which are expensive when it comes to fibre. The protocol that is spoken on the links is vendor specific, therefore debugging is not possible on customer side. In this case a management tool would only be able to show the desired behaviour, not the detailed actual state.

Conclusion
We improved the quality and the availability of the network infrastructure in the last decades widely. The unreliable two-tier architecture was replaced by a more reliable three-tier approach.

We also eliminated every single point of failure:

- A redundant core architecture was created.
- We use nowadays a redundant ISP connection.
- Stateful firewalling is provided at the provider edge.

At the moment we offer reliant and state of the art service for a distributed university.

We are able to offer access to the internal networks from every internet connection. The used procedure is error-prone and requires a lot of human resources. This should be improved via automation and consistency checks.

Through the constantly growing complexity, better debugging tools and methods are needed.

3 Further Work

We will continue to work on the reliability, maintainability and scalability of the network infrastructure. To this end we are creating a management solution, that is able to support debugging, monitoring, and automation.

Currently we are planning to restructure our network with optical fibre and multiprotocol label switching (MPLS). With these technique we will be able to centralize the firewalling and use the scaling and managing abilities from a firewall cluster system. We also will be able to use virtual private LAN service, this is needed e.g. to connect multiple data centers that need layer 2 connectivity.

References

1. Köhler, S., Binzenhöfer, A.: MPLS traffic engineering in OSPF networks-a combined approach. In: Proceedings of 18th International Teletraffic Congress ITC, pp. 21–30 (2003)
2. RFC-3031: Multiprotocol Label Switching Architecture. https://tools.ietf.org/html/rfc3031 (2001)

3. RFC-2328: OSPF Version 2. https://tools.ietf.org/html/rfc2328 (1998)
4. Vanbever, L., Vissichio, S.: Enabling SDN in old school networks with software-controlled Routing Protocols. In: Open Networking Summit (ONS) research track (2014)
5. IEEE Std 802.1D TM - 2004: IEEE Standard for Local and metropolitan area networks Media Access Control (MAC) Bridges (2004)

Interconnection Network: Design Space Exploration of Network for Supercomputers

Kentaro Sano

Abstract This chapter reports design space exploration for an interconnection network subsystem in a supercomputer targeting memory-intensive applications, which had been conducted in a feasibility study for the next-generation supercomputers. Supercomputers composed of many nodes with high-performance processors also require a high performance network so that computing performance can be scaled up to the entire system. In our design space exploration, we discuss feasibility of networks with various topologies based on 64×64 high-radix crossbar switches with the assumption of a system overview and available technologies by 2018. Preliminary evaluation suggests that a full fat tree network and a fat-tree-torus hybrid network are promising in terms of a network diameter and the number of links, in comparison with the other topologies including a 3D or 5D torus and a dragonfly. In detailed evaluation, we estimate communication performance, cost, and power consumption for the two promising networks. The obtained results show that the hybrid network can have shorter latency for local point-to-point communication than the full fat tree network while both the networks can have comparable performance for collective communications.

1 Introduction

Due to continuing demand for higher performance computation, supercomputers have been developed challenging a problem to bring the highest performance with available computing technologies. Currently, after entering Peta-Flops era, the high-performance computing (HPC) community shares the common next goal to achieve Exa-Flops computing performance with a single system. While several research and development projects have been conducted in different countries, Japan had a two-year feasibility study (FS) project, where three selected teams had studied feasibility of the next-generation supercomputers based on different architectures. Tohoku university, NEC corporation, and JAMSTEC (Japan agency

K. Sano (✉)
Graduate School of Information Sciences, Tohoku University, 6-6-01 Aramaki Aza Aoba, Aoba-ku, Sendai 980-8579, Japan
e-mail: kentah@caero.mech.tohoku.ac.jp

M.M. Resch et al. (eds.), *Sustained Simulation Performance 2014*,
DOI 10.1007/978-3-319-10626-7_13

for marine-earth science and technology) joined one of the teams which was led by Prof. Kobayashi with Tohoku university. The team had studied a supercomputer for memory-intensive applications, which is based on a vector processor architecture.

In this chapter, we introduce our design space exploration of a network subsystem for the supercomputer, which was done by a network-subsystem working group in the team. Supercomputers, which are massively parallel computers, have been getting more and more computing nodes with higher performance than before. Especially, in a supercomputer for memory-intensive applications, a vector processor with significantly wide memory bandwidth brings very high computing performance to each node. Such a system also requires a high performance network, that is, a high-throughput and low-latency network. Or the system fails into performance saturation caused by overhead of communication among nodes. Thus we have to find the best solution with available technologies to meet requirements for both computing and networking.

Simultaneously, we need to take care of target applications to design a really useful and efficient system. Some applications may mainly require global, or collective communications. On the other hand, there also exist other types of applications that mainly use local point-to-point communications, for example, ones based on a domain decomposition approach. As requirements depend on applications, we have to consider not only the technology constraint, but also communication patterns demanded by applications. We should also take care of other aspects, such as usability, performance robustness, and scalability of a system.

The goal of our working group is to find an interconnection network subsystem suitable for the next-generation supercomputers, by exploring design space with application demands and technology constrains. Especially we are aiming at a small-diameter and low-latency network using high-radix switches, which can also be good at local communications. We mainly evaluate performance, cost, and power in design space exploration while we also discuss usability.

This chapter is organized as follows. Section 2 introduces assumption for the design space exploration. Section 3 describes preliminary comparison to find the final candidates for the detailed evaluation. Section 4 shows the detailed evaluation for a full fat-tree network and a fat-tree-torus (FTT) hybrid network. Finally, section "Conclusions" gives conclusions and future work.

2 Assumption for Design Space Exploration

In exploring design space of an interconnection network, we made assumption for an entire system and technologies that will be available by 2018. The assumption for a system is summarized in Fig. 1. We assumed that the system has up to 65,536 SMP (simultaneously multi-processing) nodes, while we finally evaluated a system with 25,600 nodes. Each node has four vector processors for SMP with a shared memory. For an interconnection network, we assumed four planes of the same network for the 4-processor SMP configuration.

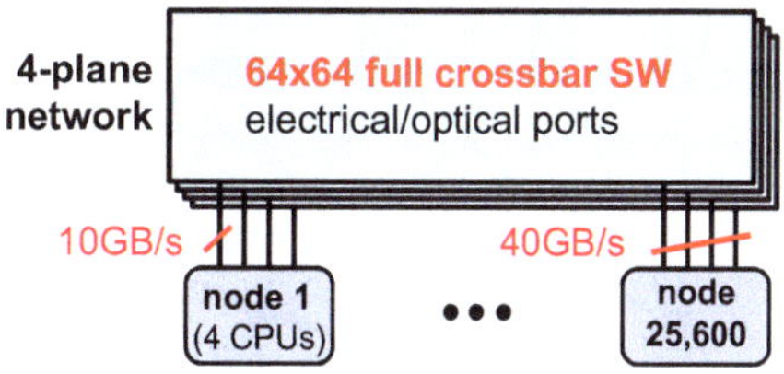

Fig. 1 Overview of a target system.

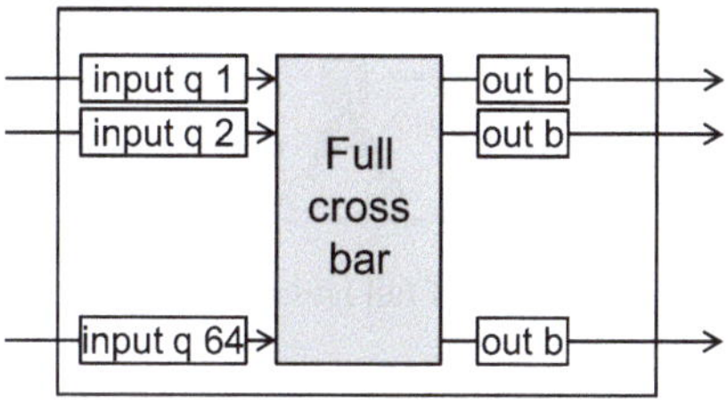

Fig. 2 64 × 64 crossbar switch

The network is composed of crossbar switches. Considering device technology that will be available by 2018 based on the current technology trend, such as Infiniband roadmap [4], we assumed that a 64 × 64 full crossbar switch is available with an electrical and optical link of 10 GB/s, which is a net bandwidth for data transfer after applying error-correction encoding in the physical layer. By using four links per node, the injection bandwidth is set 40 GB/s per node. Figure 2 shows a block diagram of the switch. The switch is based on the technology of virtual cut-through and four virtual channels. We assumed that the switch has a four-banked full crossbar with static priority, and a switching latency of 100 cycles at an operational frequency of 1 GHz. We use these numbers as a basis of our design space exploration.

3 Preliminary Comparison Among Possible Topologies

At the beginning of the design space exploration, we conducted preliminary evaluation with qualitative comparison among network topologies for 65,536 nodes. We selected a full fat tree, a n-D torus, a dragonfly, and a fat-tree-torus (FTT) hybrid topologies for the preliminary evaluation, some of which are typical ones used for supercomputers.

Figure 3 shows the full fat tree, which is well-known and has been used in the NEC SX supercomputer series. For example, the vector supercomputer of the earth simulator [5] used an interconnection network with the full fat tree. Figure 4 shows the n-D torus, where n is usually 2, 3, 4, 5, and 6. The torus is another well-known network topology, and has been used in IBM BlueGene/L [1], Cray XT4 [3], and Fujitsu K-computer [2]. Figure 5 is a relatively new network topology, which was proposed in 2008 [6]. The dragonfly is a hierarchical topology, composed of an intra-group network and an inter-group network. Connecting multiple switches

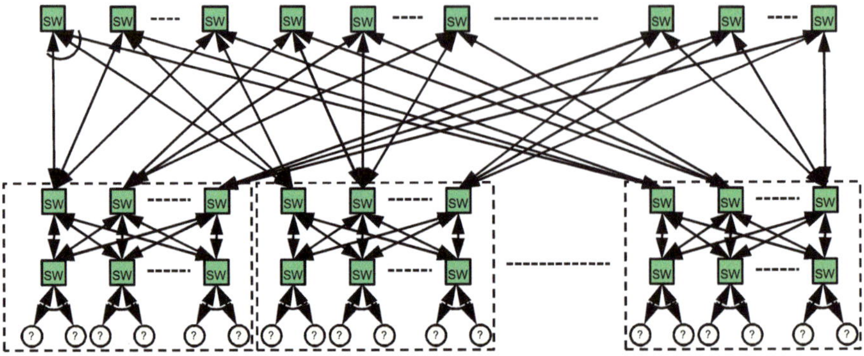

Fig. 3 Network of full fat tree

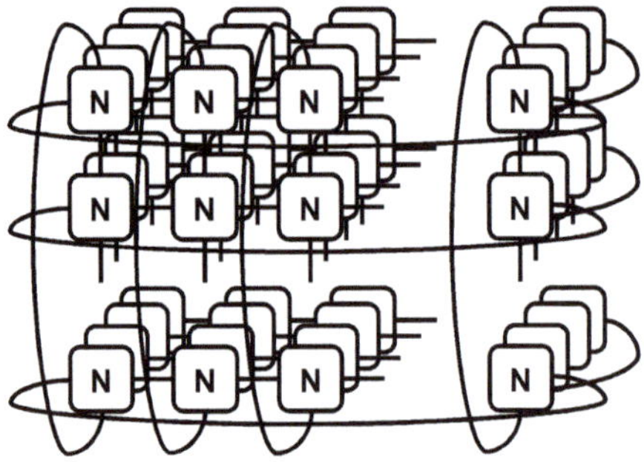

Fig. 4 Network of n-D Torus

makes a pseudo high-radix switch as a group. In our evaluation, we adopted the 3D all-to-all connection to form a group with 16,384 nodes, and 1D all-to-all connection of the 16 groups for 65,536 nodes. The dragonfly network was used in Cray Cascade [8].

In addition to these network topologies that have been used in the existing supercomputers, we introduced and evaluated another hierarchical topology, called a fat-tree-torus (FTT) hybrid shown in Fig. 6. As described later, the full fat tree is disadvantageous for cable length and delay especially for a larger number of nodes, while it has an advantage for the small number of hops. In the FTT hybrid network, we restrict the growth of cable length by restricting a full fat tree to connection of the 256 nodes in each island, and we expand a network by connecting island with a 2D torus network. The reason why a 2D torus network is used is because the cabinets of nodes have a physically planar layout on a floor of a building. If we can install the cabinets or nodes in 3D, a 3D torus would be better solution. To sufficiently give a bandwidth between islands, we connect the adjacent islands with 128 links, which corresponds to the half of the total injection bandwidth per island. As shown in Fig. 6, each island has $4 \times 128 = 512$ links for the four directions. We assumed that the networks with the full fat tree, the dragonfly, and the FTT hybrid are composed

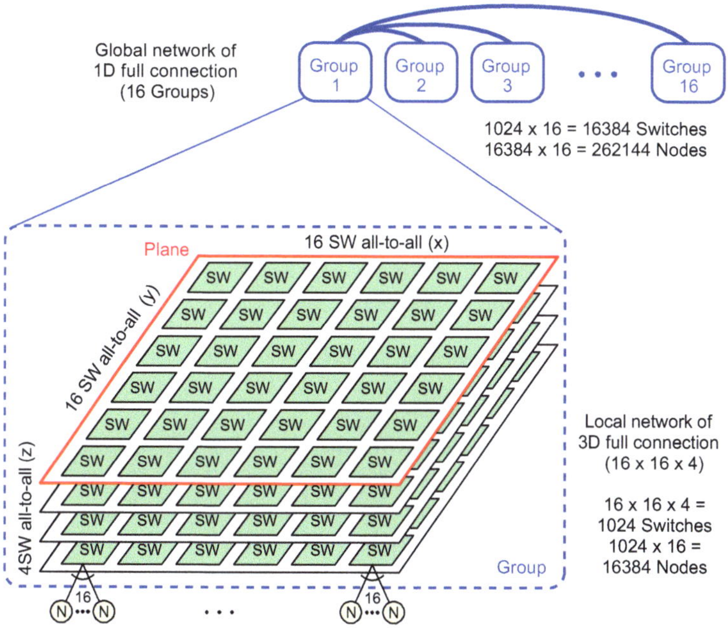

Fig. 5 Network of dragonfly

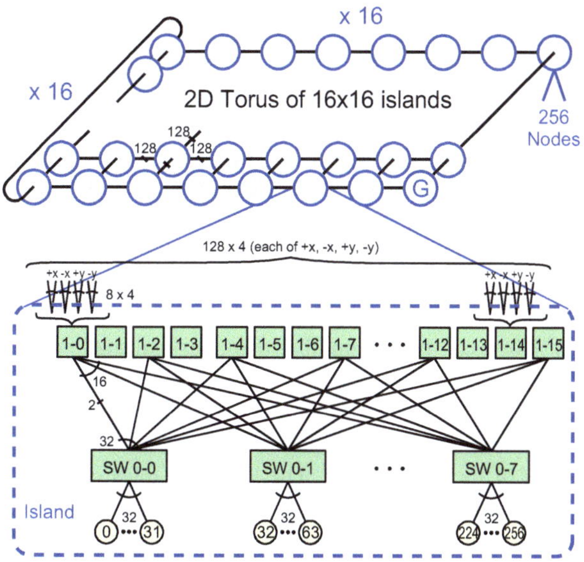

Fig. 6 Network of fat-tree torus (FTT) hybrid

Topology	Full fat-tree	3D torus	5D torus	Dragonfly	FTT hybrid
Nodes	65,536	65,536	65,536	65,536	65,536
Organization	3 stages 64 x 32 x 32	64 x 32 x 32	16x8x8x8x8	all-to-all (1D 16, 2D 16x16)	2D torus (16x16) full fat tree (256)
Node injection BW [GB/s]			10		
Bisection BW [TB/s]	320	20	80	160	40
Min to Max hops	2 to 6	1 to 63	1 to 23	2 to 5	2 to 20
Min to Max delay [ns]	100 to 500	100 to 6,300	100 to 2300	100 to 400	100 to 1,900
Links	196,608	196,608	1,310,720	468,736	196,608
Switches	5,120	within nodes	within nodes	4,096	6,144

Fig. 7 Comparison among four selected topologies

of the 64×64 full crossbar switches, while the network with n-D torus uses switches implemented within each node.

Figure 7 shows analytical comparison for the selected topologies, where $n = 3$ or 5 for n-D torus. Please note that we use only a single network plane for 65,536 nodes, resulting in the node-injection bandwidth of 10 GB/s. This table is clearly telling that the 3D torus has a too large diameter, which means that a lot of hops are required for communication between most distant nodes. Although we can reduce the diameter by increasing the dimension to 5, then the number of required links gets very large. Too many links cause a severe problem for cost and power consumption.

On the other hand, the dragonfly network has a small diameter similar to the full fat tree. However it also requires too many links, which give a weak point to the dragonfly. The full fat tree has very good numbers for bisection bandwidth, network diameter, and the number of links. As a result of the above discussion, we selected the full fat tree as a candidate topology for the detailed evaluation.

However, we also noticed that the fat tree has a potential problem when we consider physical implementation with a lot of nodes. That is a cable length and delay. Figure 8 shows the three-stage organization of the full fat tree for 65,536 nodes. We introduce islands, each of which is a group of 1,024 nodes connected by the two-stage switches. The islands are connected by the top switches, referred to as spine switches. Thanks to the high-radix switch, we can reach any nodes within six hops. However the big number of nodes can make cables connected to the spine switches very long. Since cabinets with nodes and switches installed are physically distributed over the floor, there exists such a pair of nodes that requires very long cables in a communication path between them. The red arrows in Fig. 8 depict an example of the long cables. The longest cables cause the longest delay because signal propagation speed is about 5 ns/m. As discussed in the next section, a layout for tens of thousand of nodes can give tens of meters to a cable with a delay of hundred nano seconds, which can be longer than the switching delay. This is a weakness of a fat tree network.

The FTT hybrid network allows us to have a moderate number of hops at most, which is smaller than that of the torus, with an acceptable number of links similarly

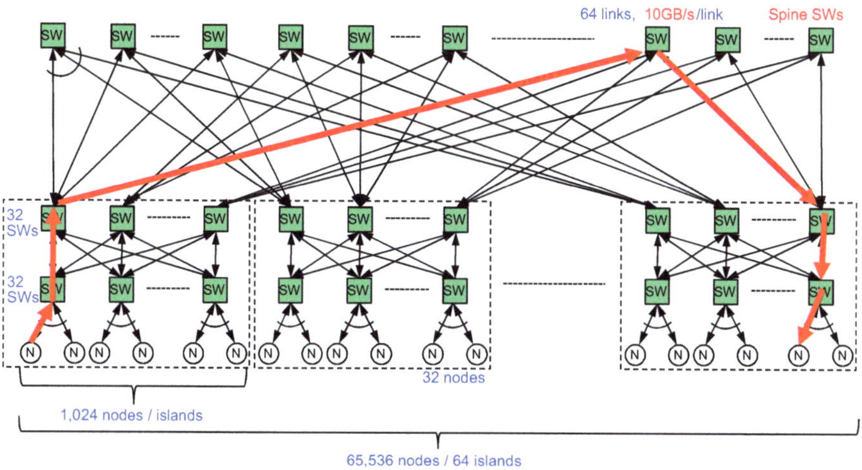

Fig. 8 Three-stage full fat tree for 65,536 nodes. The *red arrows* depict the longest cables between the most distant nodes in different islands

	Features	Diameter	# of Links	Note
Fat tree	General-puropse, high usability	very good	good	High cable delay?
Low-D torus	Good cost performance, extendability	bad	good	-
High-D torus		good	bad	-
Dragonfly	Pseudo high-radix NW	very good	bad	-
FTT-hybrid	Combination of fat tree and torus	good	good	Low cable delay?

Fig. 9 Summary of preliminary evaluation for four topologies

to the full fat tree. The number of links for the FTT hybrid is smaller than that of the 5D torus and the dragonfly. Thus the FTT hybrid can be another promising candidate. The hybrid is expected to be advantageous for cable lengths because the 2D torus topology that globally connects the islands requires only short cables for adjacent cabinets. By using the folding technique, we can avoid the longest cable for end-to-end connection that should be added to a 2D mesh connection.

The result of the preliminary evaluation is summarized in Fig. 9. The full fat tree network, which is general-purpose with high usability, is very good at diameter and good at the number of links. The low or high-dimensional torus network, which has good cost performance and extendability, is bad at diameter or the number of links, respectively. The dragonfly network, which is a pseudo high-radix network, is bad at the number of links while it is very good at diameter. Our FTT hybrid network, which is of combination of fat tree and 2D torus, is good at both diameter and the number of links. Based on them, we made a choice of the full fat tree and the FTT hybrid for the detailed evaluation considering implementation and floor layout.

4 Detailed Evaluation

We evaluated the selected two topologies, the full fat tree and the FTT hybrid, in detail by considering implementation with actual system parameters. First of all, we estimated physical installation of nodes and switches. Although we can not describe all of them in this manuscript, for example, we considered such a floor plan as shown in Fig. 10 for the system with 25,600 nodes connected by the FTT hybrid network. Then we estimated lengths and delay of cables that are used in each network organization. We also counted longer links as optical cables while shorter links are counted as electrical cables. We used 100 ns for switching delay of a single crossbar switch, including the transceiver and receiver delay. We also estimated prices of switches chips, cables, and transceivers and receivers for electrical and optical links. We assumed that the crossbar switch chips consume most of power in the interconnection network subsystem.

Figure 11 shows the result of the detailed evaluation for the 4-plane network to connect 25,600 nodes. The full fat tree requires a slightly smaller number of switches than those for the FTT hybrid. Both the topologies use the same number of cables for each of optical and electrical connections. The price of the FTT hybrid network is estimated to be 5 % higher than that of the full fat tree mainly because of the difference in the number of switch chips. The power consumption of the FTT hybrid network is 8 % higher than that of the full fat tree. This is also mainly due to

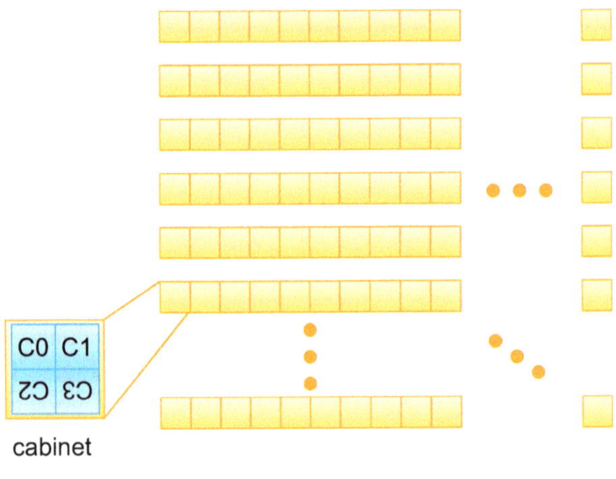

Fig. 10 Layout example for FTT hybrid network

	Fat tree	Hybrid
Number of SWs	8,448	9,600
Optical links	102,400	102,400
Electrical links	102,400	102,400
Price	1	1.05
Power	1	1.08
Diameter(hops)	6	12
Inter-island End-to-end delay (most distant nodes)	1,410 ns	1,820 ns (595 ns for adjacent nodes in Torus)

Fig. 11 Evaluation results for the two candidate network topologies

the more switch chips used for the FTT hybrid network. However, these difference in cost and power is negligible considering the total cost and power consumption of the entire system.

The diameter of the network, that is the maximum number of hops, is 6 for the full fat tree while it is 12 for the FTT hybrid. However, difference in the longest delay of end-to-end communication between most distant nodes is not so big. The inter-island end-to-end delay is 1,410 ns including the switching delay and the cable delay, while it ranges from 595 to 1,820 ns for the FTT hybrid. The maximum delay of the FTT hybrid is only 23 % longer than that of the full fat tree in spite of the doubled diameter. The actual inter-island end-to-end delay depends on which island is the destination. If we choose the adjacent island in the 2D global torus network, it is reduced to only 595 ns, which is much shorter than that of the full fat tree. This means that some local point-to-point communication can be mapped well to the FTT hybrid by using these adjacent short connections.

However, the global torus network in the FTT hybrid can degrade collective communication performance, such as broadcast and allreduce communication implemented in the MPI library [7]. To evaluate the degradation, we conducted network simulation for a single plane of a full fat tree or an FTT hybrid connecting 32,768 nodes. We developed a network simulator and described MPI-like message passing scenario to be evaluated. In the simulation, we consider the message-passing setup time on a CPU, the time of frame transmission through cables including transceiver and receiver delay, between the switch chips, and collision delay at the output ports of the switch chip. Please note that the setup time per message passing is relatively very long, which was estimated with the measurement on an existing vector supercomputer by NEC. If we have collisions of frames at switches, they cause longer communication time.

Figure 12 shows the results. The result of Sendrecv is a time of communication where two distant nodes simultaneously send a single message to each other. The x-axis shows the size of a message to be sent. Since no collision occurs in this

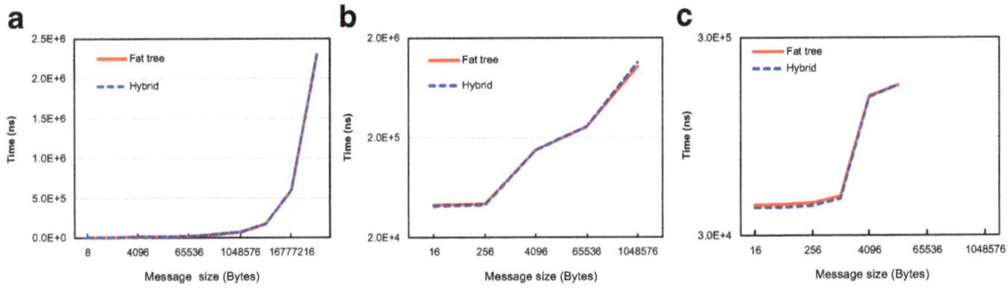

Fig. 12 Simulation results(**a**) Sendrecv (**b**) Bcast (**c**) Allreduce

case and the setup time is relatively long, the two topologies give almost the same communication time to Sendrecv. We used a binary tree algorithm to describe message passing for the broadcast. Regardless of the 2D global torus that is not good at collective communication, the FTT hybrid achieved comparable broadcast time to that by the full fat tree. We used a recursive doubling algorithm to describe the allreduce communication. Figure 12 shows that the FTT hybrid gives slightly longer communication time than that by the full fat tree. These comparable performance of the FTT hybrid is because the global communication on the 2D torus is successfully described by using available links in parallel as much as possible.

Conclusions

This cheaper reports our design space exploration for an interconnection network subsystem in a supercomputer targeting memory-intensive applications. With the assumption of the system overview and available technologies by 2018, we discuss feasibility of networks with various topologies based on 64 × 64 high-radix crossbar switches. In the preliminary evaluation, we made sure that the full fat tree and the FTT hybrid are promising in terms of a network diameter and the number of links, in comparison with the other topologies including the 3D or 5D torus and the dragonfly. Although the FTT hybrid has a doubled diameter, the longest cable can be much shorter than that of the full fat tree network. In the detailed evaluation, we discuss the cost, the power, the end-to-end latency, and communication time for the full fat tree and the FTT hybrid. The results show that both the networks have comparable performance for collective communications including broadcast and all gather, while the full fat tree achieves slightly better performance. On the other hand, the FTT hybrid can have shorter latency for local point-to-point communication between adjacent islands due to its shorter cables.

However, the FTT hybrid has difficulty in mapping processes to nodes so that the best communication performance is obtained. In terms of usability,

(continued)

the full fat tree is easy to get good communication performance because all the nodes of the fat tree can be used in the same way. Based on this, we treat the full fat tree as a primary candidate for further study. We introduce some hardware support for the full fat tree to reduce its setup time per message-passing.

Acknowledgements I appreciate the support by the Tohoku-NEC-JAMSTEC team members, and specially thank the members of the network subsystem working group: Mr. Takashi Hagiwara, Dr. Yasushi Kanoh, Dr. Sintaro Momose with NEC, Prof. Go Hasegawa and Prof. Shinichi Arakawa with Osaka univerisity, and Prof. Hiroyuki Takizawa with Tohoku university.

References

1. Adiga, N.R., Almási, G., Almasi, G.S., Aridor, Y., Barik, R., Beece, D., Bellofatto, R., Bhanot, G., Bickford, R., Blumrich, M., et al.: An overview of the bluegene/l supercomputer. In: Proceedings of Supercomputing, pp. 60–60 (2002)
2. Ajima, Y., Sumimoto, S., Shimizu, T.: Tofu: A 6D mesh/torus interconnect for exascale computers. Computer **42**(11), 36–40 (2009)
3. Alam, S.R., Barrett, R.F., Eisenbach, M., Fahey, M.R., Hartman-Baker, R., Kuehn, J.A., Poole, S.W., Sankaran, R., Worley, P.H.: The cray xt4 quad-core: a first look. In: Proceedings of the 50th Cray User Group (2008)
4. I.T. Association: Infiniband. http://www.infinibandta.org/
5. Habata, S., Yokokawa, M., Kitawaki, S.: The earth simulator system. NEC Res. Dev. **44**(1), 21–26 (2003)
6. Kim, J., Dally, W.J., Scott, S., Abts, D.: Technology-driven, highly-scalable dragonfly topology. SIGARCH Comput. Architect. News **36**(3), 77–88 (2008)
7. Snir, M., Otto, S., Huss-Lederman, S., Walker, D., Dongarra, J.: MPI: A Message-Passing Interface Standard. The MIT Press, London (1996)
8. Faanes, G., Bataineh, A., Roweth, D., Court, T., Froese, E., Alverson, B., Johnson, T., Kopnick, J., Higgins, M., Reinhard, J.: Cray cascade: a scalable HPC system based on a dragonfly network. In: Proceedings of the International Conference on High Performance Computing, Networking, Storage and Analysis (SC'12), Salt Lake City, pp. 103:1–103:9. IEEE Computer Society Press, Los Alamitos (2012). http://dl.acm.org/citation.cfm?id=2388996.2389136

Part III
Computational Approach Towards Engineering and Multi-Physics Applications

Experiences in Developing HPC Software with Portable Efficiency

Daniel Friedrich Harlacher, Harald Klimach, and Sabine Roller

Abstract This contribution is concerned with software engineering for HPC applications in the field of engineering with a focus on CFD. Our goals are general usability, maintainability, portability and of course high performance and high scalability. Usability is a key property to any application as it sets the bounds on the efficiency the user can work with it. Therefore, we focus on this aspect and show how it is entangled with the other goals.

Engineering deals with the design of technical devices and their fine tuning. Increasingly efficient devices require highly sophisticated models. This does not only entail a high resolution but also multiple physical phenomena, complex geometries and different length and time scales to be considered. Simulation software in the field of engineering needs to address these requirements, including high performance to handle the large computational demands. This results in sophisticated applications, which are deployed in an increasingly complex machine environment and need to combine fast execution with involved implementations for large models.

We share in this article experiences we made in the development of a general mesh based tool chain for engineering applications named *APES* and its deployment on HPC machines. Specifically we address on the one hand some software engineering concepts, how they are applied, where they can be neglected and which of them we deem especially important in the HPC context. On the other hand we point to shortcomings in the machine installations that require workarounds and break software engineering principles.

1 Introduction

The capability to represent more and more complex models with the steadily increasing computational power available to us results also in ever more complex simulation software. While this trend results in larger and more complex simulation

D.F. Harlacher (✉) • H. Klimach • S. Roller
Simulation Techniques and Scientific Computing, University of Siegen, Siegen, Germany
e-mail: daniel.harlacher@uni-siegen.de; harald.klimach@uni-siegen.de; sabine.roller@uni-siegen.de

M.M. Resch et al. (eds.), *Sustained Simulation Performance 2014*,
DOI 10.1007/978-3-319-10626-7_14

software deployed on supercomputers, it does not touch the observation that the largest fraction of computing time is only spent in a small part of the code [5]. High performance computing is mainly concerned with this small part of the code that takes most of the time during execution. The art of software engineering in this environment boils down to embedding these parts in the larger application framework without sacrificing performance on the one side and code design on the other. On top of these requirements software developers for supercomputing applications today face the need for highly parallel and distributed methods and implementations as increasing computational power is mostly gained by increased distributed parallelism. While this parallelism is generally taken into account for the compute intensive parts, it often is neglected in the other parts. However, distributed parallelism needs to be deployed from the first step to the last to ensure scalability of the complete simulation workflow and the capability to deal with huge data sets.

Needs of the HPC community with respect to software engineering in the development of scientific applications has been analyzed by various authors [2, 12] and guidelines have for example been proposed by Baxter et al. [3]. They offer 5 practices for a successful scientific software development. We follow those suggestions to the largest part, but as pointed out in [2] by Basili et al., a main objective to software applied in HPC is the minimization of the time to solution. They include here the time spent on software development itself, but from our perspective another important factor for this time to solution is the usability of the applications, the ease of deployment and the analysis of obtained results. Usability has to be respected by all tools in the workflow of the simulation.

How efficiently a software can be deployed not only depends on the sustained performance achieved on given computing machines, but also on the usability exhibited by it. Only if a user can efficiently deploy a software for his applications with varying scenarios it will be used to solve a large number of problem sets. Besides improving program execution it thus is also important to keep a focus on the usability of the application. Often in scientific applications for engineering, users are also the developers and there is only a small emphasize put on general usability. To broaden the userbase of supercomputing devices, this ought to change and more effort should be put into the usability of software designed for high performance computing. Even though usability has to be understood here under the constraints of the specialized user community. For example HPC user often prefer automizable scripts over graphical interfaces.

In our simulation framework *APES* [11] we aim for these goals and provide a flexible user configuration with high sustained performance [15] and highly scalable (~100,000 processes) implementations for all steps required in the workflow. The need for a tool chain providing parallelism in every step involved in large scale simulations was claimed for example by Tu et al. [13]. With this work we present some experiences gained in the development of *APES* and show the software engineering deployed.

Some goals aimed for in HPC coincide with general software engineering goals, like keeping up maintainability, portability and usability of the code. Portability even has an increased importance in HPC, as the systems are so diverse, requiring recompilations for each system to obtain the best possible performance. The requirements on portability are also in a sense more strict in the HPC context, as it includes portable performance, that is the application should be capable to execute efficiently on largely varying computing systems without further ado. Due to the need for recompilations, when switching the system and environment this goal is also directly related to the usability. If the application is not installed by the site administrators, the user has to do the compilation himself and thus, the ease of compilation on various systems becomes a question of usability. Compiling applications is often complicated by shortcomings in the software stack available on the different HPC installations. Such issues can not be completely overcome by the application design in itself and results in road blocks for some software engineering principles. Especially the reusage of code by relying on libraries is weakened by this, as an extensive dependency on third party libraries diminishes the portability. In our experience there is likely at least some shortcoming or bug in one part of an installed software stack. To avoid the failure of the complete application due to such a problem, the application has to minimize its usage of third party libraries. Even then bugs might arise in the essential system environment, like compilers and the MPI libraries, that need to be worked around specifically.

Another part contributing to usability, not yet mentioned is the documentation of the software. Documentation rarely can be extensive enough, and with frequently changing code as seen in scientific applications, there is the need to always adapt documentation with the code changes. Various tools and methods exist to address the individual goals in the software engineering process and we will exemplarily present some, that we found useful and that are deployed in the *APES* framework.

2 Building Blocks in HPC Software Design

After introducing the goals and software, we are concerned with, we now move on and highlight some of the building blocks we use in the development of our software stack in *APES*.

2.1 Implementation Language

We are dealing with large sets of numbers and target high sustained performance. A suitable and established language in this respect is Fortran. It is ideally suited for the small fractions of the code that consume more than 90 % of the computing cycles. Our main development in *APES* therefore is done in modern Fortran. With the Fortran 90 standard, modules where introduced that also enhanced the suitability of Fortran in terms of software engineering concepts, as it allowed for breaking the code down into individual components with well defined interfaces. Object oriented features were more specifically introduced in the Fortran 2003 standard, that also introduced an interoperability layer for C libraries. By this evolution of code features in the language, Fortran now allows a natural blend of highly optimized code with strongly design oriented code parts. In conclusion we see Fortran also as a suitable candidate for the remaining >90 % of code, where performance is not the most predominant factor.

We use Fortran modules to separate the large and complex software framework into smaller and maintainable encapsulated components. Each module is to be as self contained as possible and provide all the necessary information and methods to describe a specific entity, resembling an object. In the largest part of the code, we extensively make use of new Fortran features like derived data types, array syntax and dynamic memory objects, while in the kernel parts we try to stick to basic features. This helps the compilers to optimize those parts and achieve a high sustained performance. It should be noted, that sometimes software design principles have to be broken in the embedding code in order to allow for highly efficient kernels. Sometimes design principles have to be abandoned to work around compiler and library short comings. For example, the OpenMP standard offers the possibility to collapse nested rectangular loops. However, the Cray compiler failed with this in our implementation during compilation, while the Intel compiler silently produced wrong code, though the validity of the code was approved by GCC, which worked as expected. The manual collapsing of those loops results in code duplications due to the need to manually compute the collapsed indices.

To create a convenient environment for the deployment of the actual simulation software, we also make use of some Python scripts. Those scripts take care of building, testing and also running the applications. They provide the ties to bind together the individual steps in a simulation and ease the deployment.

2.2 Portability

For HPC applications a high portability is essential due to the large range of computing systems and environments. In our experience Fortran provides an excellent portability if the program sticks to the standard. Fortran compilers are available on all supercomputing platforms, however the environments in terms of libraries and supported features might greatly vary. On top of the mere compiling it thus becomes necessary to deal with proper configuration of the software in those various environments. Yet, compilation in itself should be an easy process, that can be repeated without large effort on various systems. A tool providing a configuration infrastructure and automatic resolution of Fortran dependencies is offered by *Waf* [9]. This build script is written in Python, self contained and usually distributed along with the application. The definition of the build process itself is a simple Python script and thus flexible enough to deal even with complicated build or configuration setups.

We use it to find the location of libraries to link against, test for features offered by the compiler or MPI libraries and provide appropriate workarounds where required parts are not present in the environment. Also known issues in the installed software stack can be addressed by this configuration tool. By explicitly checking for required features instead of relying on certain version numbers, the configuration gets both, more robust and wider applicable. *Waf* also offers the possibility to maintain different build variants with independent configurations.

To maintain performance across various machines, we offer some compile time configurations to adapt to the architecture of the executing machine. One example here is a loop length limit to fit parts of the computation into the cache, another is the memory layout of the state representation in the simulation. These are compile time options so the compiler might exploit the knowledge about the adapted properties and produce better optimized code for example by vectorization.

2.3 Ease of Use

Besides the simple installation of the applications on different HPC systems, a flexible runtime configuration system is required to provide a convenient deployment of the application. As HPC systems are usually run in batch operation, a text based user interface is the least cumbersome and most efficient path to communicate with the user. This text based configuration then has to be readable and understandable for humans and yet powerful enough to allow for complex settings. HPC users are comfortable with the usage of the shell and scripting allows for automation that is often necessary to do parameter studies. Scripting thus, seems to be an ideal candidate for the user interface in this environment as long as portability is not broken. We think the Lua scripting language [6] is apt for this purpose and especially well suited for HPC, due to its high portability. Lua is written in pure ANSI-C without further mandatory dependencies, only a few kilobytes large and released under the permissive MIT license. With these properties it is possible to distribute the whole library along with the application and compile it on the target machines together with the actual application. To interface with the C library offered by Lua, we use the Fortran library *Aotus* [7] to encapsulate it and provide a more Fortran specific interface.

The usage of a full-fledged scripting language like Lua for the user interface enables a more flexible usage of the application. Users can adapt configurations to their needs. It becomes possible to exploit logic and loops in the configuration. Even functions might be defined, that then can be evaluated later on by the application at runtime, this is especially useful for expressions, that are regularly subject to change from simulation to simulation, like initial and boundary conditions or source terms. It easily allows the annotation of configuration settings by comments and the reuse of one setting in another one, allowing for algebraic expressions. At the same time using scripting in the environment of the software to glue together interacting parts and ease the setup of many similar simulation runs, helps the productivity further. These scripting tools provide also the option to seamlessly generate jobs to be run in batch systems of HPC systems, lowering the effort needed for this task without introducing a graphical user interface. The benefit of scripting and automated actions is completed by tools to extract graphs and visualization files from simulation results.

Another building block enabling others to actually use the software, is the documentation. A major issue with documentation is not only to write it in the first place but also to keep it up to date as wrong documentation is worse than none. Thankfully, there are several tools available to ease the maintenance of the documentation and keeping them in sync as proposed by Knuth in the concept of literate programming [8]. The nice thing about this concept is that it does not affect the performance at all and can be used extensively throughout the code. We use Doxygen [14] to generate documentation out of comments. Again, documentation generation should be simple, just as compilation and we integrate the generation of documentation in our build scripts to ensure this. Additionally, the documentation is generated on a nightly basis and published on a web server to make access to up to date documentation even more convenient.

Both, documentation and ease of deployment are the major building blocks to broaden the user base of supercomputing applications beyond the original developer group.

2.4 Maintaining a Scientific HPC Application

As already pointed out, the software is broken down into components using the Fortran module facilities. Another level of separation is obtained by the organization of mostly independent larger tasks in separate executables and libraries. They are maintained in the distributed version control system Mercurial [10] as independent repositories. We found this a suitable option for our needs as it merely depends on Python and thus does not introduce further dependencies to our development environment. Project management and coordination utilizes Redmine [1] that is also used to host central copies of the software code repositories. Redmine offers a ticketing system to keep track of open issues and tasks, which enables a more transparent collaboration and keeps track of the development visible to all team members. Unit tests are integrated into the build process with *Waf* and are by default executed upon each compilation. Though we do not fully deploy a test driven development system [4], we aim for a high coverage by tests and found them tremendously useful in keeping the projects functional. In addition we deploy a regression check, that executes a set of test cases on a nightly basis and informs about success or failure via e-mail.

By allowing the development in branches within the Mercurial system, developers are free to pursue their own agendas and play with new ideas, while the continuous testing maintains a fairly stable default version that serves as a common ground for the whole team. This offers a nice compromise between flexibility and stability.

Conclusions
We showed some of the software engineering tools we use in the development of our HPC application framework *APES* and highlighted the methods we deem most important. Even though high performance is the dominating goal in the software suite, general software engineering concepts can be deployed in a large part of the development. Besides the primary goal of high sustained execution performance, we emphasize the ease of use to reduce the effort needed to obtain scientific results from simulations. A major tool to achieve this are in our opinion supporting scripting techniques for configuration and execution of the simulations. Portability of code and performance is an especially important trait for software in the HPC landscape due to the large diversity of systems, but also needs to cope with the increasing complexity in the system software stack.

Acknowledgements The authors would like to thank the Gauss Centre for Supercomputing (GCS) which provided us with the necessary resources on different HPC systems.

References

1. Andriy L.: Mastering Redmine. Packt Publishing, Birmingham (2013)
2. Basili, V., Carver, J., Cruzes, D., Hochstein, L., Hollingsworth, J., Shull, F., Zelkowitz, M.: Understanding the high-performance-computing community: a software engineer's perspective. IEEE Softw. **25**(4), 29–36 (2008)
3. Baxter, S.M., Day, S.W., Fetrow, J.S., Reisinger, S.J.: Scientific software development is not an oxymoron. PLoS Comput. Biol. **2**(9), e87 (2006)
4. Erdogmus, H., Morisio, M., Torchiano, M.: On the effectiveness of the test-first approach to programming. IEEE Trans. Softw. Eng. **31**(3), 226–237 (2005)
5. Fernandes, E., Barbosa, V., Ramos, F.: Instruction usage and the memory gap problem. In: Proceedings 14th Symposium on Computer Architecture and High Performance Computing, 2002, pp. 169–175 (2002)
6. Ierusalimschy, R., Figueiredo, L.H.D., Celes, W.: Lua 5.1 Reference Manual. Roberto Ierusalimschy (2006)
7. Klimach, H.: Advanced options in tables and universal scripting. https://bitbucket.org/apesteam/aotus/wiki/Home (2011). Accessed 30 July 2013
8. Knuth, D.: Literate programming. Center for the Study of Language and Information, Stanford (1992)
9. Nagy, T.: The Waf Book. http://waf.googlecode.com/svn/docs/wafbook/single.html (2010). Accessed 30 July 2013
10. O'Sullivan, B.: Mercurial: The Definitive Guide. O'Reilly, Sebastopol (2009)
11. Roller, S., Bernsdorf, J., Klimach, H., Hasert, M., Harlacher, D., Cakircali, M., Zimny, S., Masilamani, K., Didinger, L., Zudrop, J.: An adaptable simulation framework based on a linearized octree. In: Resch, M., Wang, X., Bez, W., Focht, E., Kobayashi, H., Roller, S. (eds.) High Performance Computing on Vector Systems 2011, pp. 93–105. Springer, Berlin/Heidelberg (2012)

12. Segal, J.: When software engineers met research scientists: a case study. Empir. Softw. Eng. **10**(4), 517–536 (2005)
13. Tu, T., O'Hallaron, D.R., Ghattas, O.: Scalable parallel octree meshing for TeraScale applications. In: Proceedings of the 2005 ACM/IEEE conference on Supercomputing, SC '05, pp. 4–19. IEEE Computer Society, Washington, DC (2005). ACM ID: 1105769
14. van Heesch, D.: Doxygen manual. http://www.stack.nl/~dimitri/doxygen/manual.html (2010). Accessed 30 July 2013
15. Zudrop, J., Klimach, H., Hasert, M., Masilamani, K., Roller, S.: A fully distributed CFD framework for massively parallel systems. In: Cray User Group 2012, Stuttgart (2012)

Petascale Computations for Large-Scale Atomic and Molecular Collisions

Brendan M. McLaughlin and Connor P. Ballance

Abstract Petaflop architectures are currently being utilized efficiently to perform large scale computations in Atomic, Molecular and Optical Collisions. We solve the Schrödinger or Dirac equation for the appropriate collision problem using the R-matrix or R-matrix with pseudo-states approach. We briefly outline the parallel methodology used and implemented for the current suite of Breit-Pauli and DARC codes. Various examples are shown of our theoretical results compared with those obtained from Synchrotron Radiation facilities and from Satellite observations. We also indicate future directions and implementation of the R-matrix codes on emerging GPU architectures.

1 Introduction

Our research efforts continue to focus on the development of computational methods to solve the Schrödinger and Dirac equations for atomic and molecular collision processes. Access to leadership-class computers allows us to benchmark our theoretical solutions against dedicated collision experiments at synchrotron facilities such as the Advanced Light Source (ALS), Astrid II, BESSY II, SOLEIL and Petra III and to provide atomic and molecular data for ongoing research in laboratory and astrophysical plasma science. In order to have direct comparisons with experiment, semi-relativistic or fully relativistic computations, involving a large number of target-coupled states are required to achieve spectroscopic accuracy. These computations could not be even attempted without access to HPC resources such as those available at leadership computational centers in Europe and the USA.

B.M. McLaughlin (✉)
Centre for Theoretical Atomic, Molecular and Optical Physics (CTAMOP), School of Mathematics & Physics, Queen's University, The David Bates Building, 7 College Park, Belfast BT7 1NN, UK
e-mail: b.mclaughlin@qub.ac.uk

C.P. Ballance
Department of Physics, Auburn University, 206 Allison Laboratory, Auburn, AL 36849, USA
e-mail: ballance@physics.auburn.edu

M.M. Resch et al. (eds.), *Sustained Simulation Performance 2014*,
DOI 10.1007/978-3-319-10626-7_15

We use the R-matrix, R-matrix with pseudo-states (RMPS) method to solve the Schrödinger and Dirac equations for atomic and molecular collision processes.

Satellites such as *Chandra* and *XMM-Newton* are currently providing a wealth of X-ray spectra on many astronomical objects, but a serious lack of adequate atomic data, particularly in the K-shell energy range, impedes the interpretation of these spectra. Spectroscopy in the soft X-ray region (0.5–4.5 nm), including K-shell transitions of singly and multiply charged ionic forms of atomic elements such as Be, B, C, N, O, Ne, S and Si, as well as L-shell transitions of Fe and Ni, provides a valuable probe of the extreme environments in astrophysical sources such as active galactic nuclei (AGN's), X-ray binary systems, and cataclysmic variables [1–3]. For example, K-shell photoabsorption cross sections for the carbon isonuclear sequence have been used to model the Chandra X-ray absorption spectrum of the bright blazar Mkn 421 [4].

The motivation for our work is multi-fold; (a) astrophysical Applications [5], (b) fusion and plasma modelling, JET, ITER, (c) fundamental interest and (d) support of experimental measurements and Satellite observations. In the case of heavy atomic systems [6, 7], little atomic data exists and our work provides results for new frontiers on the application of the R-matrix; Breit-Pauli and DARC parallel suite of codes. The current state of parallelism for these codes is outlined and some indication of new directions being explored with emerging architectures is presented. These highly efficient codes are widely applicable to the support of present experiments being performed at synchrotron radiations facilities such as; ALS, ASTRID II, SOLEIL, PETRA III, BESSY II. Various examples of large scale calculations are presented to illustrate the predictive nature of the method.

The main question asked of any method is, how do we deal with the many body problem? In our case we use first principle methods (ab initio) to solve our dynamical equations of motion. Ab initio methods provides highly accurate, reliable atomic and molecular data (using state-of-the-art techniques) for solving the Schroedinger and Dirac equation. The R-matrix non-perturbative method is used to model accurately a wide variety of atomic, molecular and optical processes such as; electron impact ionization (EII), electron impact excitation (EIE), single and double photoionization and inner-shell X-ray processes. The R-matrix method provides highly accurate cross sections and rates used as input for astrophysical modelling codes such as; CLOUDY, CHIANTI, AtomDB, XSTAR that are necessary for interpreting experiment/satellite observations of astrophysical objects and fusion and plasma modeling for JET and ITER.

After the Iron Project one needs to proceed with the Trans-Iron Peak Elements; (Trans-Iron Peak Project: TIPP), since various UV spectra of hot post-AGB stars exhibit unidentified lines. Some of these lines may stem from highly ionized (VI–VIII) metals that are enriched by the s-process from atomic elements such as: Ge, Se, Ba, Sr, Y, Pb. Determination of abundances would be most useful for these elements. However lines from such ions, are hitherto not identified as the main problem is precise wavelengths are required for unambiguous line identifications. Some examples are, Ge, Pb, ..., in cooler white dwarfs and sub-dwarfs, however abundances are useless in these cases because they are strongly affected by diffusion. Palmeri and co-workers have initiated the DESIRE project (http://w3.umons.ac.be/astro/desire.shtml) to address the current deficiency of atomic data for such heavy atomic elements.

2 Parallel R-matrix Photoionization

The use of massively parallel architectures allows one to attempt calculations which previously could not have been addressed. This approach enables large scale relativistic calculations for trans-iron elements such as; Kr-ions, Xe-ions [6]. It allows one to provide atomic data in the absence of experiment and takes advantage of the Linear algebra libraries available on most architectures. We fill in our sea of ignorance i.e. provide data on atomic elements where none have previously existed. The present approach has the capability to cater for Hamiltonian matrices in excess of 250 K $\times$ 250 K. Examples will be shown in the following sections for both valence and inner-shell photoionization for systems of prime interest to astrophysics and for more complex species necessary for plasma modelling in fusion tokamaks.

The development of the dipole codes, benefit from similar modifications and developments made to the existing excitation R-matrix codes. In this case all the eigenvectors from a pair of dipole allowed symmetries are required for bound-free dipole matrix formation. Every dipole matrix pair is carried out concurrently with groups of processors assigned to an individual dipole. The method is applicable to photoionization, dielectronic-recombination or radiation damped excitation and now reduces to the time taken for a single dipole formation. The method so far implemented on various parallel architectures has the capacity to cater for photoionization calculations involving 500–1,000 levels. This dramatically improves (a) the residual ion structure, (b) ionization potential, (c) resonance structure and (d) can deal with scattering channels in excess of 4,000.

3 Scalability

As regards to the scalability of these R-matrix codes, we find from experience on a variety of petaflop machines that various modules within this suite of codes scale very well, upwards to 100,000 cores. In practical calculations for cross sections on various systems it is necessary to perform fine energy resolution of resonance features ($\sim 10^{-8}$ (Ry) $\sim$1.36 meV) which is observed in photoionization cross sections. This involves many (6–30 million) incident photon energies, vital when comparing with high precision measurements such as those performed on Xe^+ at the Advanced Light Source synchrotron radiation facility in Berkeley, California, USA where energy resolutions of 4–9 meV FWHM are achieved.

The formation of many real symmetric matrices (Hamiltonians), typically 60–150 K, requires anywhere from 10–500 Gb of storage. The diagonalization of each

Table 1 B II 249-states, 400 coupled channels, 409,600 energy points running on increasing number of cores

CRAY-XE6	PSTGB0FDAMP	PSTGB0FDAMP	PSTGB0FDAMP
CPU cores	Absolute timing (s)	Speed up factor	Total core hours
1024	584.19	1.0000	166.1155
2048	430.80	1.3584	245.0077
4096	223.08	2.6183	253.8154
8192	149.70	3.9018	340.6506

The results are from module PSTGB0FDAMP for photoionization cross-section calculations of the B II ion carried out on HOPPER the Cray XE6 at NERSC. Results are presented indicating the speed up factor with increasing number of CPU cores and the total number of core hours

matrix, from which *every* eigenvalue and *every* eigenvector is required is achieved through use of the ScaLapack package. In particular routines: **pdsyevx** and **pdsyevd**, where preference is given to the latter, as it ensures orthogonality between all eigenvectors. In typical collision calculations matrices vary in size from $2\,\mathrm{K} \times 2\,\mathrm{K}$ to $200\,\mathrm{K} \times 200\,\mathrm{K}$, depending on the complexity of the atomic target. The formation of the continuum-continuum part of the $\mathrm{N} + 1$ electron Hamiltonian is the most time consuming. Therefore if there are several thousand scattering channels then there are $[nchan \times (nchan + 1)/2]$ matrix blocks. Each block represents a partial wave and each subgroup reads a single Hamiltonian and diagonalizes it in parallel, concurrently with each other. So there is endless scalability. R-matrix close-coupling calculations are therefore reduced to the time required for a single partial wave.

In Table 1 we show details of test runs for the outer region module PSTBF0DAMP for K-shell Photoionization of B II using 249-coupled states with 400 coupled channels and for 409,600 energy points and an increasing number of CPU cores. A factor of 4 speed up is achieved by using up to 8,192 cores. The computations were carried out on the Cray-XE6 (Hopper) at NERSC. Note, for actual production runs, timings would be a factor of 10 larger, as one would require a mesh of 4,096,000 energy points to fully resolve the resonances features observed in the spectrum. We present the timings for core sizes varying from 1,024 to 8,192 again for B II K-shell photoionization in its ground state. The computations were done with the outer region module PSTGBF0DAMP for 249-states and 400-coupled channels.

The main work horse in our linear algebra code is the ScaLAPACK libraries. The goals of the ScaLAPACK project are the same as those of LAPACK; Efficiency (to run as fast as possible), Scalability (as the problem size grows so do the numbers of processors grow), Realiability (including error bounds), Flexibility (so users can construct new routines from well-designed parts) and Ease of Use (by making the interface to LAPACK and ScaLAPACK look as similar as possible). Many of these goals, particularly portability are aided by developing and promoting standards, especially for low level communication and computation routines.

Parallel I/O issues have been addressed for the photoionization module, PST-GBF0DAMP, as several large passing files (ranging from 10 to $-150\,\mathrm{Gb}$) need to be read during runtime. These files are SEQUENTIAL FORTRAN BINARY files. For large-scale computations using the module PSTGBF0DAMP, file names have a typical size, e.g. Hamiltonian file: H.DAT (10–150 Gb), dipole files: D00, D01, D02, DO3, …, DNN (2–50 Gb). Traditionally, all processors read these files together. For a small numbers of processors, say less than 500 processors and file sizes (<5 Gb) both on the Hopper Cray XE6 (NERSC) and the Kraken Cray XT5 (National Institute for Computational Science, NICS) architectures, the jobs perform well, with the I/O time being $<10\,\%$ of the total wall clock time. However, as the number of processors increases, i.e. 5,000–6,000 processors, the I/O percentage of wall clock time is typically 33 % pre-striping. This I/O issue has been addressed using efficient striping of the data files. The Lustre file system is made up of an underlying set of I/O servers and disks called Object Storage Servers (OSSs) and Object Storage Targets (OSTs) respectively. A file is said to be striped when read and write operations access multiple OST's concurrently. File striping is used to increase the I/O performance of our codes writing or reading from multiple OST's simultaneously increases dramatically the available I/O bandwidth. We have investigated the following I/O issues along with the use of striping files.

1. The H.DAT file, unlike the D00, D01, ... files is read sequentially from start to finish by all processors, however by allowing ONLY processor 0 to read, then distribute (i.e. MPI_BCAST) the information to the remaining processors is a better solution. This technique however cannot be implemented for the dipole (DXX) files, as each processor is reading different parts of the file, and different amounts of information at each read. Table 2 shows the minimal increase in performance for module PSTGBF0DAMP for the large-scale photoionization cross-section calculations on Ca II using this approach.
2. Each DXX file holds enough information to describe the photoionization process from the ground state and ALL excited states of an atom/ion. If the user/scientist requires only a small subset of all possible photoionization cross sections, (i.e. ground-state and first metastable) then the DXX files could be REDUCED in size.

Table 2 Ca II, 513-states, 1815 coupled channels, 6,000 energy points running on 6,000 cores

CRAY-XE6	PSTGBF0DAMP[a]	PSTGBF0DAMP[b]
Ca II	Absolute timings (s)	Absolute timings (s)
6,000 cores	108.2344[c]	116.2098
Maximum coupled channels	1815	1815
D file size	6 Gb	6 Gb
H matrix size	23073	23073

The results are from module PSTGB0FDAMP for photoionization of Ca II carried out on HOPPER the Cray XE6 at NERSC

[a] With READ/MPI_BCAST

[b] Without READ/MPI_BCAST

[c] Time in seconds

3. Modification of the directory structure on the LUSTRE filesystem, so that the large files may be efficiently STRIPED increases the I/O performance of our codes. We have carried out various striping investigations in collaboration with a computational scientist consultant Vince Betro, at NICS, to optimize the 'best striping' based on file size. Furthermore detailed testing has been done on the Cray XE6 at NERSC (Hopper), the Cray XT5 at NICS (Kraken), and on the Cray XE6 at HLRS (Hermit), in Stuttgart, Germany. Our findings are as follows. For file sizes of <1 Gb, we use the default striping of files (Hopper has a values of 2, Kraken a value of 1, Hermit a value of 4). For file sizes of between 1 and 10 Gb we use small stripling values, typically 20 gives optimal throughput depending on the system load, which is only apparent at the NICS facility. In the case of file sizes between 10 and 100 Gb a medium stripling value of 60 appears to be optimal. For files in excess of 100 Gb a striping value of 120 was used.

Optimization of the R-matrix codes on a variety of HPC architectures implements several good coding practices. We use code inlining (which reduces overhead from calling subroutines), loop unrolling, modularization, unit strides, vectorization and dynamic array passing. Highly optimized libraries such as; ScaLapack, BLAS and BLACS are used extensively in the codes. On Cray architectures, the R-matrix codes are profiled with performance utilities such as CrayPat and IPM which gives the user an indication of their scalability. Standard MPI/FORTRAN 90/95 programming is used on the world's leading-edge petaflop high performance supercomputers. Serial (legacy) code to parallel R-matrix implementation is carried out using a small subset of basic MPI commands.

4 X-ray and Inner-Shell Processes

To measure the chemical evolution of the universe and then understand its ramifications for the formation and evolution of galaxies and other structures is a major goal of astrophysics today. *HUT* and *IUE* observations have been used to study the processing and dredging that goes on during the lifetime of a star. Gas that is photoionized by a hard radiation field (AGN/quasars, IGM, cataclysmic variables, X-ray binaries, etc.…) can be stripped of its inner-shell electrons, and several phenomenon then occur which can radically alter the ionization balance of the gas. Space-based UV observations of emission and absorption lines from ions of C, N, O, S and Si in photoionized sources play an important role in addressing many of the astrophysical issues listed above.

Absolute K-shell photoionization cross sections for atomic nitrogen have been obtained from high resolution experiment at the ALS and the R-matrix with pseudo-states approach (RMPS). Due to the difficulty of creating a target of neutral atomic nitrogen, no high-resolution K-edge spectroscopy measurements have been reported for this important atom. Interplay between theory and experiment enabled identification and characterization of the strong $1s \rightarrow np$ resonance features

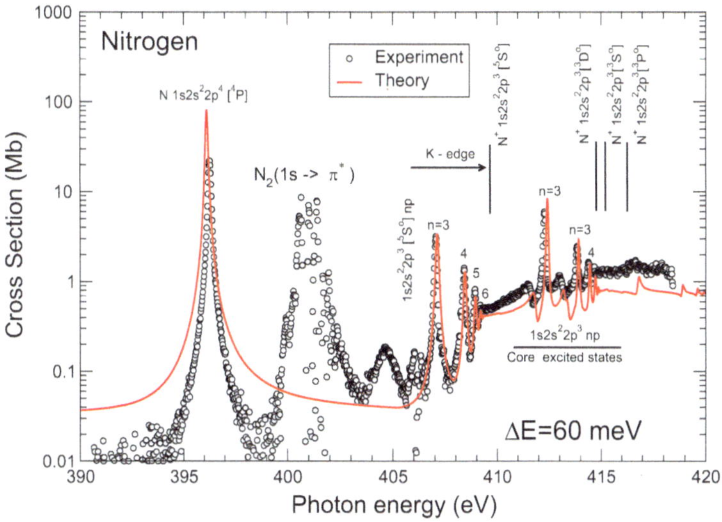

Fig. 1 Atomic-nitrogen total photoionization cross section. Theoretical results from the R-matrix with pseudo-states method (RMPS) were convoluted with a 60 meV FWHM Gaussian to simulate the ALS experiment. Experimental results include molecular components between 400 and 406 eV [8]

throughout the threshold region. An experimental value of 409.64 ± 0.02 eV was determined for the *K*-shell binding energy. Figure 1 illustrates the comparison between theory and experiment in the *K*-shell region from our recent work on this system [8].

An accurate description of the photoionization/photoabsorption of atomic oxygen is important for a number of atmospheric and astrophysical applications. Photo-absorption of atomic oxygen in the energy region below the $1s^{-1}$ threshold in X-ray spectroscopy from *Chandra* and *XMM-Newton* is observed in a variety of X-ray binary spectra. Photo-absorption cross sections determined from an R-matrix method with pseudo-states (RMPS) and high precision measurements from the Advanced Light Source (ALS) are presented in Fig. 2. High-resolution spectroscopy with $E/\Delta E \approx 4{,}250 \pm 400$ were obtained for photon energies from 520 to 555 eV at an energy resolution of 124 ± 12 meV FWHM. *K*-shell photoabsorption cross-section measurements were made on atomic oxygen at the ALS. Natural line widths Γ are extracted for the $1s^{-1}2s^22p^4(^4P)np\ ^3P^\circ$ and $1s^{-1}2s^22p^4(^2P)np\ ^3P^\circ$ Rydberg resonances series and compared with theoretical predictions. Accurate cross sections and line widths are obtained for applications in X-ray astronomy. Excellent agreement between theory and the ALS measurements is shown which will have profound implications for the modelling of X-ray spectra and spectral diagnostics. Further details can be found in our recent study on this complex [9].

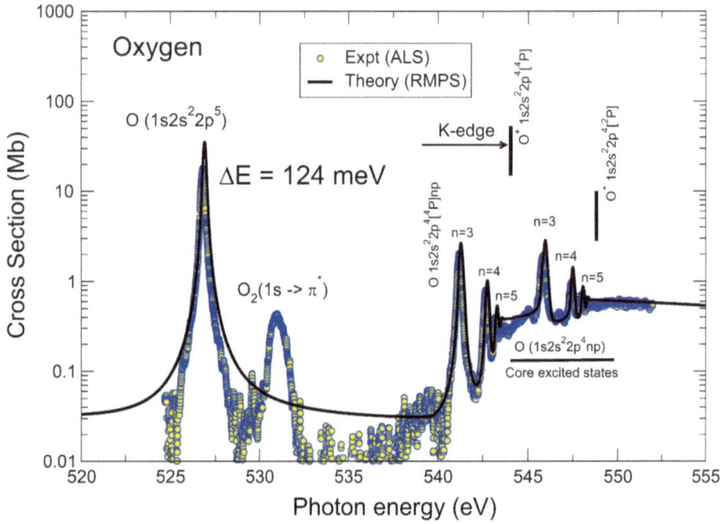

Fig. 2 Atomic oxygen photo-absorption cross sections taken at 124 meV FWHM compared with theoretical estimates. The R-matrix calculations shown are from the R-matrix with pseudo-states method (RMPS: *solid black line*, present results) convoluted with a Gaussian profile of 124 meV FWMH [9] (Colour online)

5 Heavy Atomic Systems

5.1 Kr and Xe Ions

Photoionization cross sections of heavy atomic elements, in low stages of ionization, are currently of interest both experimentally and theoretically and for applications in astrophysics. The data from such processes have many applications in planetary nebulae, where they are of use in identifying weak emission lines of n-capture elements in NGC 3242. For example, the relative abundances of Xe and Kr can be used to determine key physical characteristics of s-process nucleosynthesis, such as the neutron exposure experienced by Fe-peak seed nuclei.

Xenon and Krypton ions are also of importance in man-made plasmas such as XUV light sources for semiconductor lithography, ion thrusters for space craft propulsion and nuclear fusion plasmas. Xenon and Krypton ions have also been detected in cosmic objects, e.g., in several planetary nebulae and in the ejected envelopes of low- and intermediate-mass stars [6].

We apply the suite of DARC codes to calculate detailed photoionization (PI) cross sections on the halogen-like ions, Kr^+ and Xe^+. Photoionization (PI) cross section calculations on the Kr^+ complex were carried out retaining 326-levels in our close-coupling calculations with the Dirac-Atomic-R-matrix-Codes (DARC). In R-matrix theory, all photoionization cross section calculations require the generation of atomic orbitals based primarily on the atomic structure of the residual ion. The

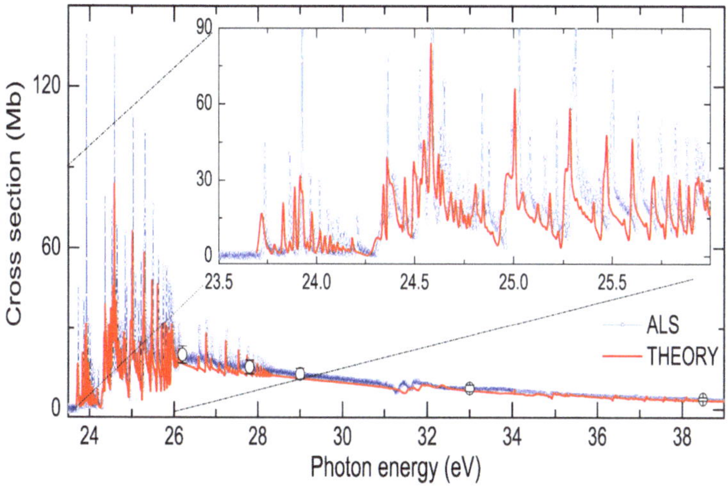

Fig. 3 An overview of measurements for the absolute single photoionization measurements of Kr^+ ions as a function of the photon energy measured with theoretical estimates for cross sections. The ALS measurements are at a nominal energy resolution of 7.5 meV and normalized to the ASTRID/SOLEIL measurements at 26.5025 eV. Theoretical results were obtained from a Dirac-Coulomb R-matrix method, convoluted with the appropriate Gaussian of FWHM and a statistical admixture of 2/3 ground and 1/3 metastable states [10]

present theoretical work for the photoionization of the Kr^+ ion employs relativistic atomic orbitals up to n = 4 generated for the residual Kr^{2+} ion, which were calculated using the extended-optimal-level (EOL) procedure within the GRASP structure code [6].

Figure 3 shows the comparison of our large-scale (326-levels) photoionization cross section calculations for Kr^+ ions with the DARC suite of codes compared with the high resolution measurements carried out at the Advanced Light Source (ALS), in Berkeley, California, USA. It is clearly see from Fig. 3 that our results obtained using the DARC codes reproduce all the fine detail resonance features in the observed spectra obtained in the ALS measurements over the photon energy range investigated [10].

Similarly photoionization cross section calculations on the Xe^+ system retained 326 levels of the residual Xe^{2+} ion in our close-coupling calculations performed with the Dirac-Atomic-R-matrix-Codes (DARC). For the Xe^+ case we have employed relativistic $n = 5$ atomic orbitals generated for the residual Xe^{2+} ion, which were obtained using the energy-average-level (EAL) procedure within the GRASP structure code on the fourteen lowest levels associated with the $5s^25p^4$, $5s5p^5$ and $5s^25p^35d^2$ configurations. For Xe^+ ions a more stringent test with recent extremely high resolution experimental measurements made at the ALS at 4 meV as illustrated in Fig. 4 indicate (in the near threshold region, apart from the n = 14 member of the Rydberg series) very good agreement giving us confidence in our theoretical results [6].

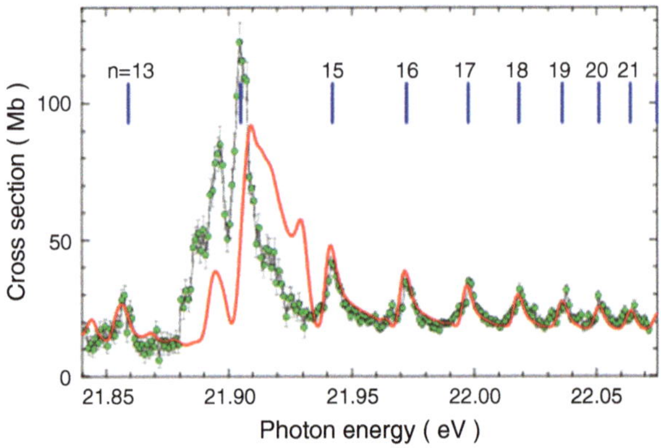

Fig. 4 Xe^{+} ALS experimental PI cross section data (*green circles*) for photon energies ranging from 21.84 to 22.08 eV at a photon energy resolution of 4 meV. Results are compared with theoretical results from a 326-level Dirac-Coulomb R-matrix calculation (*red line*) convoluted with a FWHM Gaussian of 4 meV and statistically averaged over the ground and metastable states to simulate the experimental measurements. The *bars* mark the energies of the [$5s^2 5p^4$ ($^3P_1$) nd] resonances obtained with a quantum defect of 0.16 [7]

5.2 *Tungsten (W) Ions*

The choice of materials for the plasma facing components in fusion experiments is guided by competing desirables: on the one hand the material should have a high thermal conductivity, high threshold for melting and sputtering, and low erosion rate under plasma contact, and on the other hand as a plasma impurity it should not cause excessive radiative energy loss. The default choice of material for present experiments is carbon (or graphite), however tritium is easily trapped in carbon-based walls and for that reason carbon is at present held to be unacceptable for use in a D-T fusion experiment such as ITER, under construction in Cadarache, France, or in a fusion reactor. In its place, tungsten (symbol W, atomic number 74) is the favoured material for the wall regions of highest particle and heat load in a fusion reactor vessel, with beryllium a possibility for regions of lower heat and particle load. ITER is scheduled to start operation with a W-Be-C wall for a brief initial campaign before switching to W-Be or W alone for the main D-D and D-T experimental program. In support of ITER and looking ahead to a fusion reactor the ASDEX-Upgrade tokamak now operates with an all-W wall, and at JET a full ITER-like mixed W-Be-C wall is being installed. Smaller-scale experiments involving tungsten tiles are carried out on other tokamaks. The attractiveness of tungsten is due to its high thermal conductivity, its high melting point, and its resistance to sputtering and erosion, and is in spite of a severe negative factor that as a high-Z plasma impurity tungsten does not become fully stripped of electrons and radiate copiously, so that the tolerable fraction of tungsten impurity in the plasma is at most 2×10^{-5}.

The walls of ITER are coated with tungsten which can therefore enter the fusion plasma. W ions impurities in a fusion plasma causes critical radiation loss and minuscule concentrations prevent ignition. High resolution experiments are currently available from the ALS on low ionization stages of W ions. We use the Dirac-Atomic-R-matrix-Codes (DARC) to perform large scale calculations for the single photoinization process and compare our results with experiment. These systems are an excellent test bed for the photoionization (PI) process where excellent agreement is achieved between theory and experiment providing a road-map for electron—impact excitation (EIE).

For photoionization of the W^{3+} ion of tungsten we investigated several different scattering models before settling on a 379 level model obtained from 9 configuration state functions of the residual ion. Figure 5 shows our theoretical results from this 379-level model obtained from the DARC codes compared with measurements made at the Advanced Light source. From the comparison made in Fig. 5 suitable agreement with the ALS experimental measurements is obtained when the theoretical results are statistically averaged over the fine-structure levels. Additional ions of tungsten are presently being explored and will be reported on in due course.

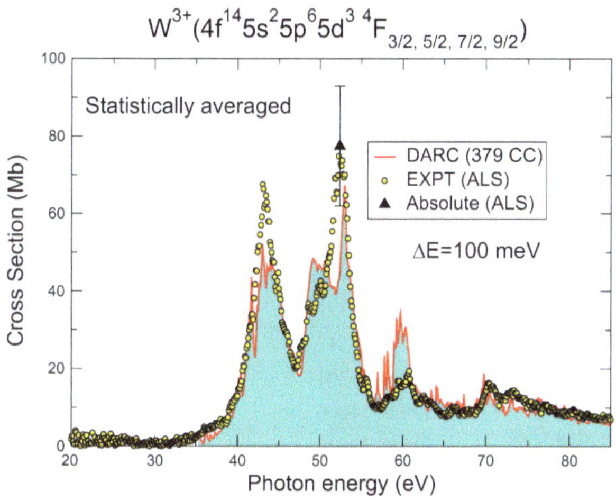

Fig. 5 Photionization of W^{3+} ions over the photon energy range 20–90 eV. Theoretical work (*solid red line*: DARC) from the 379 level approximation calculations, convoluted with a Gaussian profile of 100 meV FWHM and statistically averaged over the fine structure $J = 3/2, 5/2, 7/2$ and 9/2 levels (see text for details). The *solid circles* (*yellow*) are from the experimental measurements made at the ALS using a band width of 100 meV and the *solid triangle* (*black*) is the absolute measurement, accurate to within 20 %

6 Future Directions and Emergence of GPUS

Experience with the test bed machine DIRAC at NERSC, a 50 core GPU architecture and the Cray XK7 (TITAN) at ORNL has indicated various things. MPI/Fortran codes access the GPU (Graphical Processing Unit) through the PGI CUDA fortran compiler and an architecture supporting NVIDIA graphics cards. Those using C/C^{++} have greater flexibility. One of the issues is identifying an intensive numerical algorithm within your code that would benefit from distribution across one or more GPUS. Individually, GPUS may be slower than their equivalent CPUS but there are many more cores. One success has been the formation of the R-matrix itself, which may be expressed as a matrix multiplication that has to be carried out repeatedly which occurs in the outer region module PSTGF. The NVIDIA kepler GPUS on DIRAC outperformed the optimized LAPACK routine DGEMM by a factor of eight, though there is still room to improve the code with CUDA BLAS. Hence formation of the R-matrix

$$R_{ij} = \sum_k \frac{W_{ik} \times W_{kj}}{(E - E_k)} \tag{1}$$

is essentially a matrix multiplication

$$\mathbf{R} = \mathbf{X} \times \mathbf{Y} \tag{2}$$

where the elements of **R** are R_{ik}, **X** are $w_{ik}/(E - E_k)$ and **Y** are w_{kj}.

One of the underlying questions is, how will performance degrade as CPU/GPU ratio increases? On TITAN, the Oak Ridge National Laboratory leadership computational facility, which has a hetrogeneous environment (CPU/GPU), a range of

Table 3 Range of timings (in seconds) using DGEMM for matrix multiplication and the ratio of CPU:GPU on the Cray XK7 (TITAN) leadership computational architecture located at Oak Ridge National Laboratory

Matrix multiplication	CPU:GPU	CPU:GPU	CPU:GPU
$n \times m$	1:1	2:1	4:1
267 × 16512	0.24[a]	0.35	0.30
308 × 19201	0.72	0.52	0.52
399 × 24666	0.51	1.10	1.12
526 × 32530	0.04	0.06	0.68
132 × 8155	0.22	0.32	0.06
260 × 16102	0.90	1.21	0.33
837 × 51573	1.03	1.40	1.71
796 × 49169	3.43	1.40	1.84
1368 × 84321	3.43	4.60	–
1241 × 76646	2.73	3.70	–

The results are from module PSTGF and for electron scattering for Fe III

[a] Time in seconds

timings using DGEMM and the ratio of CPU:GPU were investigated for the outer region module, PSTGF. We investigated the case of electron scattering from the Fe III ion. Ten different matrix sizes were investigated with 16 MPI tasks, therefore the MPI process carries out 160 multiplies of each matrix. Table 3 shows the results for these timings for the increasing CPU: GPU ratio and increasing matrix size.

Acknowledgements C P Ballance was supported by US Department of Energy (DoE) grants through Auburn University. B M McLaughlin acknowledges support by the US National Science Foundation through a grant to ITAMP at the Harvard-Smithsonian Center for Astrophysics and a visiting research fellowship from Queen's University Belfast. The computational work was carried out at the National Energy Research Scientific Computing Center in Oakland, CA, USA, the Kraken XT5 facility at the National Institute for Computational Science (NICS) in Knoxville, TN, USA and at the High Performance Computing Center Stuttgart (HLRS) of the University of Stuttgart, Stuttgart, Germany. The Kraken XT5 facility is a resource of the Extreme Science and Engineering Discovery Environment (XSEDE), which is supported by National Science Foundation grant number OCI-1053575. This research also used resources of the Oak Ridge Leadership Computing Facility at the Oak Ridge National Laboratory, which is supported by the Office of Science of the U.S. Department of Energy under Contract No. DE-AC05-00OR22725.

References

1. McLaughlin, B.M.: Inner-shell photoionization, fluorescence and auger yields. In: Ferland, G., Savin, D.W. (eds.) Spectroscopic Challenges of Photoionized Plasma, Astronomical Society of the Pacific. The ASP Conference Series, vol. 247, pp. 87. San Francisco (2001)
2. Kallman, T.R.: Challenges of plasma modelling: current status and future plans. Space Sci. Rev. **157**, 177 (2010)
3. McLaughlin, B.M., Ballance, C.P.: Photoionization, fluorescence and inner-shell processes. In: McGraw-Hill (ed) McGraw-Hill Yearbook of Science and Technology, pp. 281. McGraw-Hill, New York (2013)
4. Hasoglu, M.F., Abdel, N.S.A., Gorczyca, T.W., Drake J.J., McLaughlin, B.M.: K-shell Photoabsorption Studies of the Carbon Isonuclear Sequence. Astrophys. J. **724**, 1296 (2010)
5. Covington, A.M., Aguilar, A., Covington, I.R., Hinojosa, G., Shirley, C.A., Phaneuf, R.A., Álvarez, I., Cisneros, C., Dominguez-Lopez, I, Sant'Anna, M.M., Schlachter, A.S., Ballance, C.P., McLaughlin, B.M.: Valence-shell photoionization of chlorinelike Ar^+ ions. Phys. Rev. A **84**, 013413 (2011)
6. McLaughlin, B.M., Ballance, C.P.: Photoionization cross section calculations for the halogen-like ions Kr^+ and Xe^+. J. Phys. B **45**, 085701 (2012)
7. McLaughlin, B.M., Ballance, C.P.: Photoionization Cross-Sections for the trans-iron element Se^+ from 18 to 31 eV. J. Phys. B **45**, 095202 (2012)
8. Sant'Anna, M.M., Schlachter, A.S., Öhrwall, G., Stolte, W.C., Lindle, D.W., McLaughlin, B.M.: K-shell X-ray Spectroscopy of Atomic Nitrogen. Phys. Rev. Lett. **107**, 033001 (2011)
9. McLaughlin, B.M., Ballance, C.P., Bown, K.P., Gardenghi, D.J., Stolte, W.C.: High precision K-shell photoabsorption cross sections for atomic oxygen: experiment and theory. Astrophys. J. **771**, L8 (2013)
10. Hinojosa, G., Covington, A.M., Alna'Washi, G.A., Lu, M., Phaneuf, R.A., Sant'Anna, M.M., Cisneros, C., Álvarez, I., Aguilar, A., Kilcoyne, A.L.D., Schlachter, A.S., Ballance, C.P., McLaughlin, B.M.: Valence-shell photoionization of Kr^+ ions: experiment and theory. Phys. Rev. A **86**, 063402 (2012)

FPGA-Based Scalable Custom Computing Accelerator for Computational Fluid Dynamics Based on Lattice Boltzmann Method

Kentaro Sano

Abstract This paper presents a tightly-coupled FPGA cluster for custom computing of fluid dynamics simulation, and evaluates its performance with prototype implementation. For scalable and efficient computation with a lot of FPGA accelerators, we propose an accelerator-domain network (ADN) that brings low-latency and high-speed data transfer by directly connecting FPGAs. We describe implementation of a prototype cluster node with four FPGAs, and their on-chip framework for high-speed data streaming and computing. In performance evaluation, we demonstrate that our custom computing machine for fluid dynamics computation with the lattice-Boltzmann method (LBM) exploits both temporal and spatial parallelism, and scales the performance well with the number of FPGAs. As a result, we achieved 98.8 % of the peak performance of 73.0 GFlop/s with four FPGAs.

1 Introduction

Numerical simulations such as computational fluid dynamics (CFD) are now indispensable technology in engineering and industry because their lower-cost acquisition capability of various data than experiments, for example, tests with a wind tunnel. In order to meet the growing demand for high-performance computation (HPC) of large-scale and high-resolution simulations, accelerators such as GPUs have been getting more popular due to their computational performance per cost. GPU's throughput-oriented architecture, which is different from latency-oriented microprocessors, allows a lot of computing units to perform massively-parallel computation of thousands of threads, bringing very high peak-performance. Thus a single GPU can provide great acceleration if the computing units are sufficiently utilized, and systems with GPUs have been getting popular [1], so that they appear in Top500 list [2].

K. Sano (✉)
Graduate School of Information Sciences, Tohoku University, 6-6-01 Aramaki Aza Aoba, Aoba-ku, Sendai 980-8579, Japan
e-mail: kentah@caero.mech.tohoku.ac.jp

M.M. Resch et al. (eds.), *Sustained Simulation Performance 2014*,
DOI 10.1007/978-3-319-10626-7_16

However, it is not easy to obtain high sustained performance close to the peak in computing with a lot of GPUs. This is because throughput and latency of inter-GPU communication limit performance improvement with the number of GPUs, resulting in performance saturation. In a common GPU cluster, GPUs are not directly connected to each other. They are connected to PCI-Express interfaces of host nodes, and require indirect communication via the node's network to transfer data to other GPUs of different nodes suffering from a long latency and a narrow bandwidth, while its overhead can be somewhat reduced by such a technique as GPU direct. As a result, the scalability of a multi-GPU system is limited especially for bandwidth-intensive and/or latency-sensitive computations. This is especially conspicuous in the case of *strong-scaling*, where the problem size per GPU shrinks as the number of GPUs increases, because each GPU loses computation required to hide the communication latency.

For efficient acceleration, custom computing is another promising approach with good scalability for large-scale systems [3, 4]. Designs dedicated to an individual application allow computing resources in a system to be efficiently utilized, being well balanced with a bandwidth of a network and memories. Such tailor-made computing is a feasible solution with a suitable platform of field-programmable gate array (FPGA). FPGA is a circuit-reconfigurable semiconductor device, which has been getting advanced with a lot of logic elements, DSP blocks, on-chip memories, and high-speed I/O blocks. Not only for bit-level or integer computing, but also for floating-point computing, FPGA now has computing performance of several hundreds of GFlops with thousands of floating point units. Vendors announce that the next generation of 14 nm FPGAs will bring 10 TFlops per chip. In addition to this high peak floating-point performance, FPGAs are also expected to be advantageous for bandwidth and power consumption. Various hard macros of differential I/Os for DDR3 memory controllers, PCI-Express interfaces and high-speed transceivers are integrated onto state-of-the-art FPGAs, giving very wide bandwidth to external memories and inter-FPGA connections. Relatively low operating frequency and power-aware architectures make FPGAs a power-efficient solution.

These trends have motivated many researches on floating-point operations with FPGAs [5–8] and FPGA-based custom computing for scientific applications including Lattice Boltzmann method (LBM) [9–14]. LBM is one of the methods to compute fluids based on generalized cellular automata with numerical computation [15]. In addition to the regular and coarse parallelism of LBM, it can be implemented as stream computation which is performed with data flowing through a kernel function [9]. Since these features are suitable for custom computing, there have been several reports of FPGA-based LBM computation.

Sano et al. have developed an FPGA-based LBM accelerator [9] where stream computation is performed by dedicated pipelines, and evaluated its performance and power consumption [16]. However, streaming throughput between an FPGA and a CPU limits the performance because all the computational stages of LBM are not fully streamed. In their design, particle translation and boundary computation are still performed by a CPU. Furthermore, parallel computing with multiple FPGAs was not considered.

Murtaza et al. reported LBM computation on a multi-FPGA system, called Maxwell [12]. Maxwell is a cluster of nodes, each of which has two FPGA boards connected via PCI interface. The FPGAs are directly connected to each other by Rocket I/O, but this is not used in their work. Each FPGA contains several processing elements (PEs) to compute a sub-lattice given by domain decomposition. Boundary data on an FPGA board are read to the host PC, communicated between adjacent PCs by message-passing, updated by each PC, and finally returned to the FPGA board for the next iteration. They showed that speedup does not increase sufficiently for more than 8 FPGAs because of the communication via the host network. In particular, the time for boundary-data transfer between a CPU and an FPGA has a big impact to the scalability in strong scaling where a problem size per FPGA shrinks due to domain decomposition. These results suggest that in order to scale the performance, custom computing with multiple FPGAs should take its own way to directly transfer data between FPGAs.

To this problem, our answer is a tightly-coupled FPGA-cluster where FPGAs are directly connected by an accelerator-domain network (ADN) with a low latency and a wide bandwidth. So far, we have proposed a parallel custom computing architecture for LBM (PCCAL), and showed its linear scalability with a performance model [13]. In this paper, we present:

1. Architecture and design of our tightly-coupled FPGA cluster,
2. Implementation of a prototype node of the cluster,
3. On-chip framework for high-speed data streaming and computing on FPGAs, and
4. Performance evaluation with custom computing of LBM on four FPGAs.

This paper is organized as follows. Section 2 presents our tightly-coupled FPGA cluster and the first prototype implementation. Section 3 describes custom machines for the lattice Boltzmann method to be computed on the cluster. Section 4 evaluates the performance and efficiency. Finally, section "Conclusions" gives conclusions and future work.

2 Tightly-Coupled FPGA Cluster for Scalable Custom Computing

2.1 Architecture of Tightly-Coupled FPGA Cluster

Figure 1 shows the architecture of a tightly-coupled FPGA cluster. The cluster is composed of computing nodes, each of which has multiple FPGA accelerators that are connected to the host node via PCI-Express interface. Each FPGA has its local DRAMs to store data for computation. For streaming data from memory to

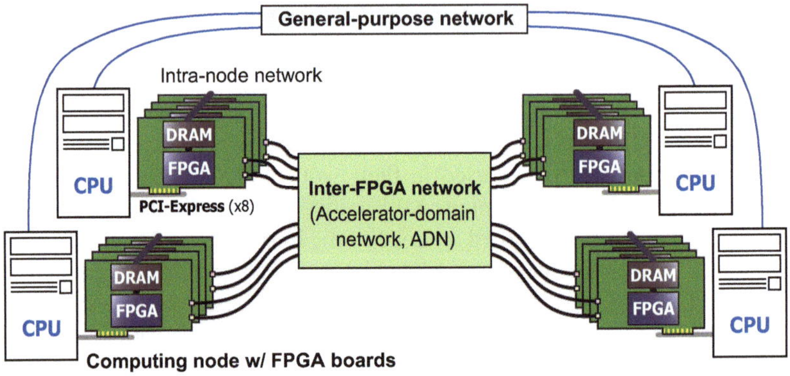

Fig. 1 Architecture of a tightly-coupled FPGA-cluster

memory, we assume two banks of memories so that read and write can be performed separately. The key feature of this architecture is the accelerator-domain network, called ADN, which directly connects FPGAs to each other. Although we can apply any network topology to ADN, here we made a choice of a 1D ring network for low latency communication, where each FPGA can transfer data directly to the adjacent FPGAs. This is because our primary application, LBM, requires local communication with adjacent FPGAs. If we need global communication among FPGAs, we can make a choice of an indirect network with switches, while FPGA-to-FPGA latency becomes relatively larger.

Recent high-end FPGAs have high-speed differential I/Os with a bandwidth higher than 10 Gbps, we can construct a high-throughput ADN for FPGAs which allows low-latency data transfer from a data-path to another data-path among FPGAs. In comparison with a conventional cluster with accelerators, we can increase the number of accelerators per node because we don't use the host-FPGA connection for inter-FPGA communication. Furthermore, the 1D ring network scales with the number of FPGAs because the we can ensure a constant bandwidth to each local connection between FPGAs.

2.2 Design and Implementation of a Cluster Node

In our research project, we have implemented a node of the tightly-coupled FPGA cluster with four DE4-230 FPGA boards designed by TERASIC Co., Ltd. [17]. Figure 2 shows a photograph of the board. The DE4-230 board has ALTERA Stratix IV EP4SGX230 FPGA [18], two DIMM slots of DDR2 memories, PCI-Express (PCIe) Gen 2.0 interface, and two high-speed mezzanine connectors (HSMCs). Stratix IV EP4SGX230 FPGA is a medium-sized device of the ALTERA's 40 nm-generation high-end FPGA series. It contains 182400 ALUTs (adaptive lookup

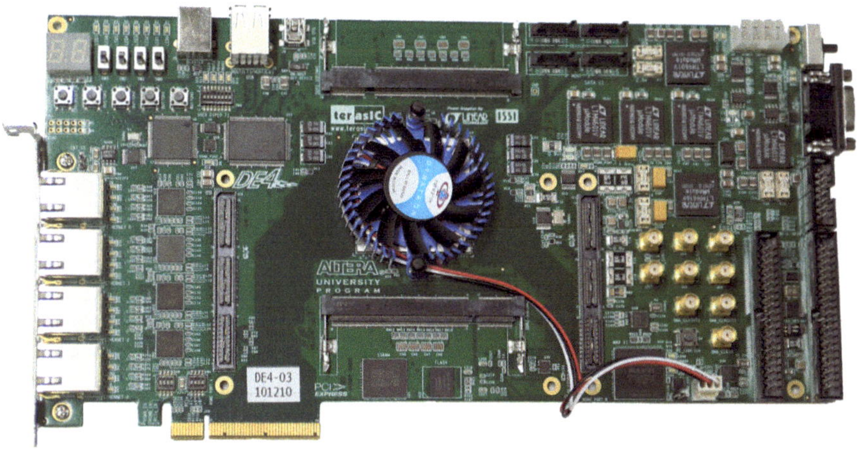

Fig. 2 Photograph of DE4-230 FPGA board [17]

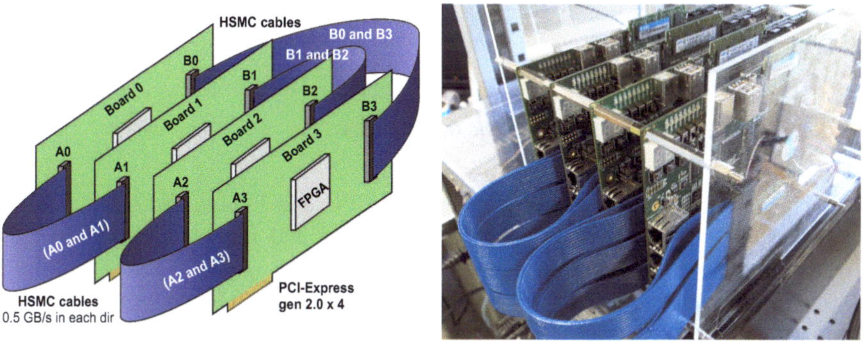

Fig. 3 Prototype cluster node with DE4-230 FPGA boards connected by a ring network

tables), block RAMS of 1828224 bytes in total, and 1288 18-bit DSP blocks, which is a hard macro of integer multiplier.

We connected four DE4-230 boards to a host PC by using PCIe Gen 2.0 with x4 lanes. We have developed a PCIe driver for Linux that provides high-speed data transfer between a CPU and one of the FPGAs, by driving a scatter-gather DMA (direct memory access) module on FPGA. As shown in Fig. 3, we made ADN of a ring network by directly connecting HSMCs of the FPGAs by their dedicated cables. We adopted single-end signaling for this connection while LVDS (low-voltage differential signaling) is also available. Single-end signaling at 125 MHz allows the cable to provide an unidirectional bandwidth of 500 MB/s at a latency of only six cycles.

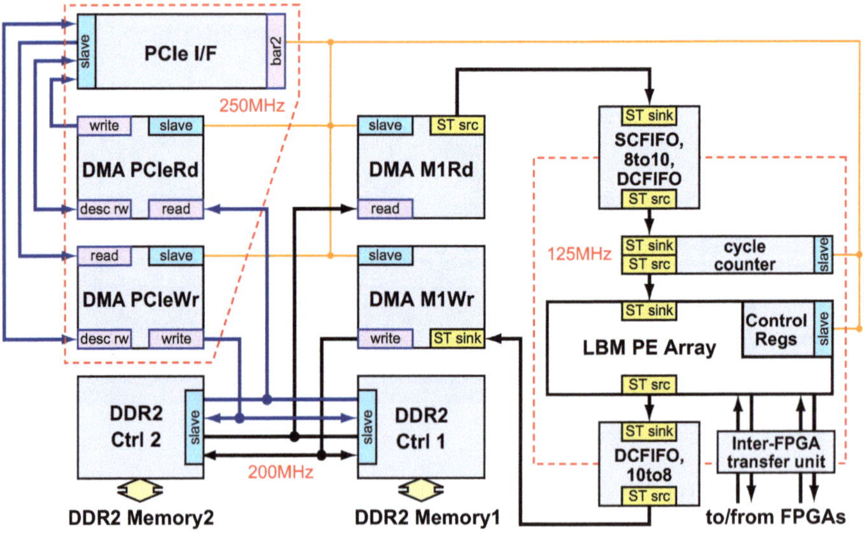

Fig. 4 On-chip system framework for high-speed data streaming

2.3 *Acceleration Framework on FPGA*

To achieve high-speed data transfer between CPU and FPGA, and a data stream between the DDR2 memories, we have designed and implemented an on-chip system framework with scatter-gather DMA (SGDMA) engines. Figure 4 shows the block diagram of the framework. We used ALTERA's system-on-programmable-chip development tool, Qsys. The modules in the framework are connected by using ALTERA's connection standards: Avalon-MM (memory mapped) and Avalon-ST (stream). The framework consists of the following modules: the PCIe interface, the two DDR2 memory controllers, the four SGDMA engines, the stream width conversion units with single-clock and dual-clock FIFOs, the cycle counter, and the user logic. Our custom computing machine for LBM is implemented with several control registers in this user logic. The PCIe, DDR2 memory controllers, SGDMA engines are available as IP cores in the Qsys module library. We have implemented the other modules by writing Verilog HDL-code so that the module interfaces satisfy the Avalon-MM or Avalon-ST.

To control the SGDMA engines, the cycle counters, and the LBM core, a program running on a host PC writes or reads control-and-status registers (CSRs) of these modules via the PCIe interface. Our Linux driver and API libraries provide these functions. For example, we can call a service function to kick SGDMA by writing its CSRs when we perform DMA transfer. The PCIe interface, the DDR2 memory controllers, and the user logic operate at different clock frequencies: 250, 200 and 125 MHz, respectively. This 125 MHz is a half clock frequency of the PCIe frequency. If we need other frequencies, we can configure PLL (phase-locked loop) modules to obtain desired clock frequencies. To transfer data between different clock-domains, we need to insert clock-crossing bridges to avoid meta-stability.

3 Case Study: Custom Computing with Lattice Boltzmann Method

3.1 Lattice Boltzmann Method

The lattice Boltzmann method (LBM) [15–21] models fluids with propagation and collision processes of fictive particles over a discrete lattice mesh. In this paper, we explain the 2D orthogonal 9-speed (2D9V) model of the lattice Boltzmann method. In 2D9V model, each cell has a distribution function f_i as shown in Fig. 5 for each of the nine particle speeds $\mathbf{c}_i = (c_{x_i}, c_{y_i})$:

$$c_{x_i} = \cos\frac{2\pi(i-1)}{8}, \quad c_{y_i} = \sin\frac{2\pi(i-1)}{8} \tag{1}$$

where $0 \le i \le 8$. Here we assume that $\Delta t = \Delta x = \Delta y = 1$ for simplicity. The propagation and collision processes on the lattice mesh are defined by the following equation:

$$f_i^{\text{new}}(\mathbf{x} + \mathbf{c}_i, t+1) = f_i(\mathbf{x}, t) - \frac{f_i(\mathbf{x}, t) - f_i^{\text{eq}}(\mathbf{x}, t)}{\tau} \tag{2}$$

where $f_i^{\text{eq}}(\mathbf{x}, t)$ is an equilibrium distribution function of particles at $\mathbf{x}$ in a lattice at time step t, τ is a single relaxation time, and Δt is a time step for particles to move to the adjacent cells. The following stages solve Eq. (2):

1. MACRO : macroscopic physical quantity calculation stage
2. EQUI : equilibrium calculation stage

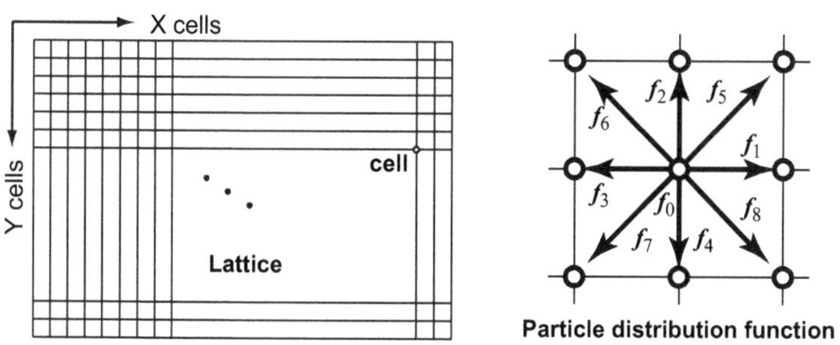

Fig. 5 2D lattice and particle-distribution functions at each cell for 2D9V model

3. COL : collision calculation stage
4. TRANS : translation stage
5. BOUND : boundary calculation stage

In MACRO stage, we compute the fluid density ρ and the fluid velocity $\mathbf{V} = (u, v)$ at $\mathbf{x}$ from the distribution functions $f_i(\mathbf{x}, t)$ by

$$\rho = \sum_i f_i(\mathbf{x}, t), \quad \rho\mathbf{V} = \sum_i \mathbf{c}_i \cdot f_i(\mathbf{x}, t). \tag{3}$$

In EQUI stage, we compute the equilibrium distribution function, $f_i^{\mathrm{eq}}(\mathbf{x}, t)$, by

$$f_i^{\mathrm{eq}} = \rho \left\{ A_i + B_i(\mathbf{c}_i \cdot \mathbf{V}) + C_i(\mathbf{c}_i \cdot \mathbf{V})^2 - D_i\mathbf{V}^2 \right\} \tag{4}$$

where A_i, B_i, C_i and D_i are constants varying according to $i = 0, 1, \ldots, 8$. In COL stage, we compute the right-hand side of Eq. (2). In TRANS stage, we propagate the particles by moving the right-hand side value of Eq. (2) at $\mathbf{x}$ to the adjacent cell at $(\mathbf{x} + \mathbf{c}_i)$ with its particle speed, $\mathbf{c}_i$.

In BOUND stage, we adjust boundary cells that require special treatment based on their conditions, e.g., for solid surfaces. We have three types of cells: fluid cells, solid cells, and outside cells. The solid-surface boundary cells, which are in contact with solid cells, are updated with the bounce-back condition [15]. The bounce-back condition simply reflects incoming particles on non-slip surfaces. For inlet/outlet boundary cells, which are in contact with outside cells, we apply a constant-pressure condition. In this condition, we calculate unknown f_i incoming to the cell with the other known f_i so that the fluid density ρ stays constant. By repeating MACRO to BOUND stages, we obtain time-dependent flow for a series of time steps.

3.2 Architecture for Stream Computation

We define stream computation for efficient custom computing of LBM:

$$\begin{aligned} \text{Inputs:} &\quad f_0, f_1, \ldots, f_8 \\ \text{Kernel:} &\quad \text{optimized Eqs. (3), (4), and (2)} \\ \text{Outputs:} &\quad f_0^{\mathrm{new}}, f_1^{\mathrm{new}}, \ldots, f_8^{\mathrm{new}} \end{aligned}$$

where the input and output correspond to a data stream generated by traversing the lattice in the x-direction first. Please note that we optimize the computation of LBM to reduce the number of floating-point operations as described in [9, 13], So far we have proposed a parallel custom-computing architecture for LBM, PCCAL, to achieve efficient and scalable computing by using multiple FPGAs with ADN [13]. Figure 6 is the overview of PCCAL to be implemented on each FPGA, which is

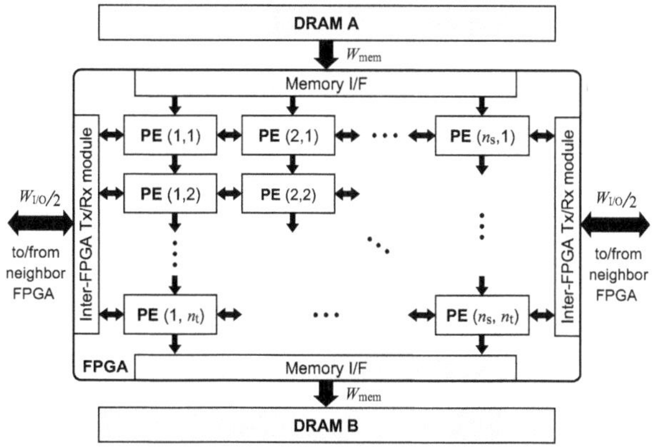

Fig. 6 Parallel computing architecture for LBM [13]

composed of the $n_s \times n_t$ PE array to exploit both spatial and temporal parallelism, the memory interfaces to stream data between external memories, and the inter-FPGA data transceiver/receiver modules to exchange the sub-lattice boundary data between the adjacent FPGAs.

We adopt 1D domain decomposition to use ($n_s N_{FPGA}$) PE columns on N_{FPGA} FPGAs for spatially parallel computation. The entire lattice of $X \times Y$ cells is partitioned into sub-lattices in x-direction. In computation, data of the sub-lattices are streamed from DRAM A and input to the columns of the PE array. In each column, the PE performs the stream computation for the allocated sub-lattice exchanging sub-lattice boundary data with the connected left and right PEs inside or outside the FPGA. Since the outputs of the first row of the PE array are considered as data streams for the next time-step, they are then input to the second row without consuming external memory bandwidth. We can increase operations executed per cycle with a constant stream-bandwidth by pipelining computations of successive time steps [14, 22]. Finally, the last row outputs the updated results to DRAM B. When streaming finishes, the streaming direction changes alternately.

3.3 PE Design for Fully-Streamed Computation

As reported in [23], we designed a processing element (PE) that performs all the stages of LBM with a stream of f_i and cell attribute. The cell attribute is a 32-bit word to give a cell information of fluid, solid, or boundary. Figure 7 shows a structure of a PE, which is composed of the following units: the MACRO, EQUI, and, COL unit (MECU) to compute cells with their floating-point pipelines, the delay unit (DU) to give delays to the cell attribute, the translation unit (TLU), and the boundary computing unit (BDU). The input stream of ten words is splitted into two streams of f_i and the cell attribute. Then, the stream of f_i is input to MECU.

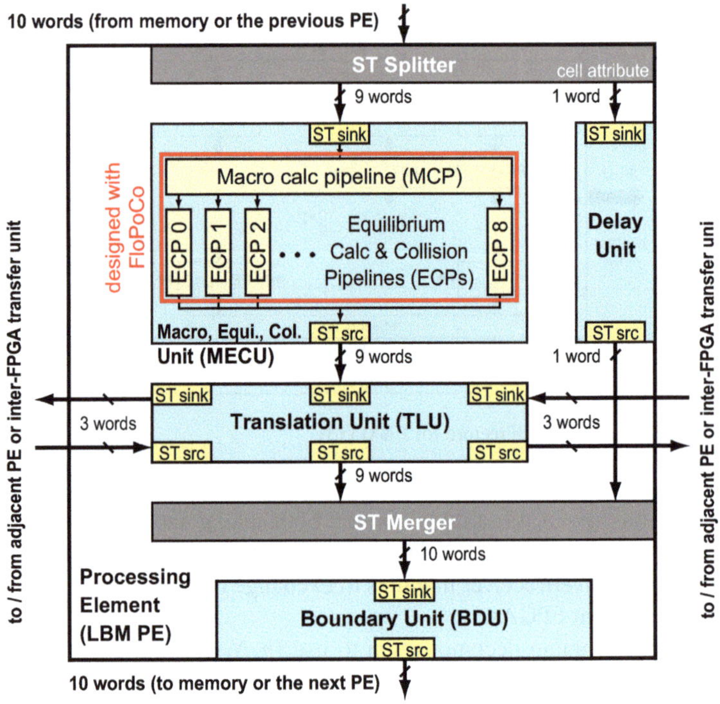

Fig. 7 Processing element [13]

MECU consists of the Macro calculation pipeline (MCP), and nine pipelines for equilibrium calculation and collision (ECPs) of f_i, respectively. MCP and ECPs have deep pipelines with a lot of single-precision operators: floating-point (FP) adders, multipliers, and a divider.

The output of MECU is streamed into TLU, and translation is performed by using shift registers as described in [13]. Moreover, TLU sends and receives data to and from the connected left and right PEs to exchange f_i with them when computing cells at sub-lattice boundaries. We design a synchronization mechanism for TLUs to control stall of the pipelines in a PE. The cell attribute from DU and the output of TLU are merged into a single stream, and then input to BDU. BDU updates cells for the bounce back condition and the constant pressure condition.

4 Performance Evaluation

4.1 Implementation of PEs

For performance evaluation, we have implemented a single PE or two cascaded PEs on each FPGA. The PE operates at 125 MHz. TLU was implemented with multi-tap shift registers for computing an $X \times Y$ lattice with $X = 360, 720, 1{,}440,\ or\ 2{,}880$. To execute LBM computation on the array, we use the SGDMA engine for PCIe to transfer the initial lattice data to DDR2 memory1 of FPGA. Then, we activate the other two SGDMA engines to stream the data between the DDR2 memories. When the stream computation finishes and the result is written into the DDR2 memory2, then we start stream computation in an opposite direction, i.e., from the DDR2 memory2 to memory1. By repeating this ping-pong streaming, the on-chip system can efficiently perform the LBM computation without accessing a host. If necessary, temporal results can be read out by the host.

4.2 *Resource Consumption*

The resource consumption of the system implemented with two PEs on one FPGA is shown in Table 1. The entire system consumes 75.4 % ALUTs (Adaptive lookup tables), 65.6 % registers, 38.0 % BRAM (block-RAM) bits, and 46.0 % 18-bit DSP blocks of those on this FPGA. The framework except the PEs consumes 20.8 % ALUTs, 22.7 % registers, 19.2 % BRAM bits, and no DSP block. The remaining resources are consumed by PEs 1 and 2. Each of PEs 1 and 2 consumes about 18 % ALUTs, 20 % registers, 2 % block-RAM bits, and 23 % DSP blocks. Although almost half of the DSP blocks are not utilized yet, a system with more than two PEs did not fit the FPGA due to ALUT and register consumption. We need to optimize resource consumption in order to implement more PEs. With the remaining BRAM bits, we estimate that the two PEs on this FPGA can compute a three-times larger lattice with $8{,}640 \times Y$ cells.

Table 1 Resource consumption of the implemented system

	ALUTs	Registers	BRAM [Kbits]	18-bit DSPs
Stratix IV EP4SGX230	182400	182400	14283	1288
Entire system	137446 (75.4 %)	119621 (65.6 %)	5433.8 (38.0 %)	592 (46.0 %)
Framework	37942 (20.8 %)	41316 (22.7 %)	2748.4 (19.2 %)	0 (0.0 %)
LBM PE array (PE x 2)	995046 (54.6 %)	78305 (42.9 %)	2685.4 (18.8 %)	592 (46.0 %)
PE 1	46889 (25.7 %)	38117 (20.9 %)	1272.4 (8.9 %)	296 (23.0 %)
PE 2	47403 (26.0 %)	38463 (21.1 %)	1404.9 (9.8 %)	296 (23.0 %)

4.3 Computational Performance

We calculate the peak Flop/s performance by multiplying the number of operators and an operational frequency. MECU contains 72 FP adders, 63 FP multipliers, and one FP divider, which operate every cycle at 125 MHz when valid data are streamed. BDU contains 7 FP adders and 3 FP multipliers. Therefore, we obtain the peak performance of a single PE, P_{PE}, as

$$P_{\mathrm{PE}} = (72 + 63 + 1 + 7 + 3) \times 0.125 = 18.25 \ \text{GFlop/s}. \tag{5}$$

The peak performance of two PEs is $2 \times P_{\mathrm{PE}} = 36.5$ GFlop/s. However, all the peak performance is not always achieved due to several overheads. In this system, the setup time for starting SGDMA, stall cycles in streaming data, and pipeline epilogue and prologue can degrade the sustained performance. We measured the time between the start and the end of SGDMA transfer by using an inline assembly code with RDTSC instruction to obtain cycle counts of x86 CPUs. By multiplying the difference in a count number with a cycle period of a CPU, we can obtain the elapsed time at high accuracy. The obtained time includes both the net computing time and the overhead.

For evaluation, we computed flow in a 2D sudden expansion channel with a solid circle of Fig. 8, by using one, two, or four FPGAs connected with an HSMC cable. We changed the lattice size from 360×120 to $5{,}760 \times 1{,}920$ in order to evaluate the pipelining overhead of a PE array. We used only 1 PE implementation for the two- and four- FPGA case because the inter-FPGA Tx/Rx module was not available yet for data exchange of two cells.

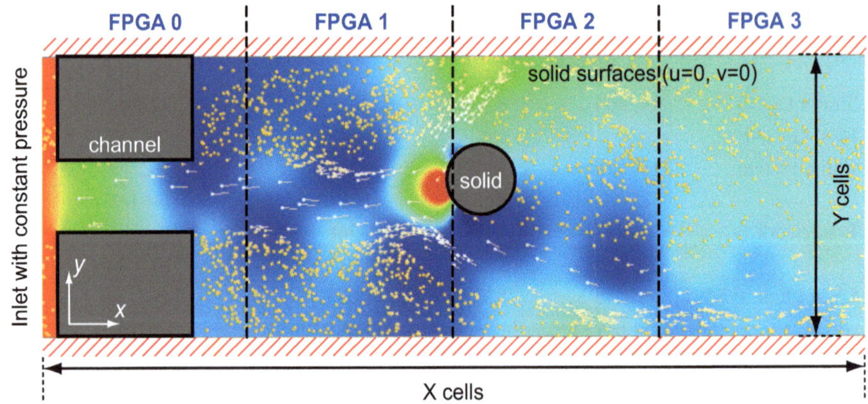

Fig. 8 Computational condition of 2D sudden-expansion channel flow with a *solid circle*. The color shows pressure, and the particles show fluid velocities. *Red* and *blue* are for high and low pressures, respectively. A Karman vortex street is made after the solid. The lattice is partitioned into four parts to be computed in parallel by four FPGAs.

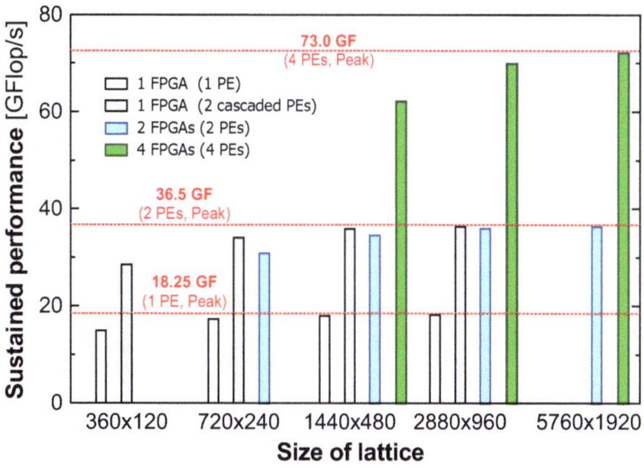

Fig. 9 Computing performance for different lattice sizes for one to four FPGAs

Figure 9 shows the computing performance of a single FPGA with one or two PEs, and two or four FPGAs with one PE each. Each bar shows GFlop/s performances obtained with measured time including the SGDMA kickoff overhead and the pipelining overhead. We observed that stall cycles are less than 1.0 % of all the cycles in almost all cases. The white bar shows the sustained performance for a single PE on one FPGA. The performance for the 360×120 lattice is 15.0 GFlop/s, which 82 % of the peak performance. This performance degradation is because of the SGDMA kickoff overhead and the pipelining overhead. As the lattice gets larger, the sustained performance increases as the overhead relatively shrinks. As a result, 18.2 GFlop/s, which is 99.7 %, was achieved for the $2{,}880 \times 960$ lattice. The similar trend is also seen in the other cases.

The light gray bar shows the sustained performance of two cascaded PEs on one FPGA. Since the depth of the pipeline is doubled in comparison with a single PE, the overhead appears larger for the 360×120 lattice. However, finally 36.3 GFlop/s, 99.5 % of the peak, was achieved with the $2{,}880 \times 960$ lattice. This demonstrates scalability by cascading PEs on an FPGA.

The light blue and green bars show the sustained performance of two FPGAs and four FPGAs with one PE on each FPGA, respectively. Since each FPGA computes a sub-lattice that is smaller than the entire lattice, the overhead is seen larger especially for a small lattice. Similarly to the other cases, increasing lattice size makes efficiency higher. The $2{,}880 \times 960$ and $5{,}760 \times 1{,}920$ lattices allow the systems to achieve 36.38 GFlop/s (99.7 % of the peak) and 72.13 GFlop/s (98.8 % of the peak), respectively. These results show that our proposed architecture has scalability with multiple FPGAs. Thus, by cascading PEs and increasing the number of FPGAs, we can exploit both the temporal and spatial parallelism to scale the performance of LBM computation with our proposed custom computing architecture for LBM.

Conclusions
This paper presents the architecture and design of the tightly-coupled FPGA cluster for custom computing of fluid dynamics simulation, and evaluates its performance with prototype implementation. The key feature of the cluster is the accelerator-domain network (ADN) that brings low-latency and high-speed direct data transfer between FPGAs. We preserve the implementation of the cluster node with four ALTERA Stratix IV FPGAs, and the on-chip framework for high-speed data streaming and computing on them. We describe the custom computing machine for fluid dynamics computation with the lattice-Boltzmann method (LBM). In performance evaluation, we demonstrate that the machine exploits both temporal and spatial parallelism of LBM computation, and scales well with the number of FPGAs. Finally, the four processing elements on four FPGAs achieves 98.8 % of the peak performance of 73.0 GFlop/s.

We are now building a larger FPGA cluster with 28 nm generation FPGAs and 10G Ethernet switch. In future work, we implement a custom computing machine of LBM for the cluster, and evaluate its performance and power consumption. We will also develop a high-level synthesis compiler to easily generate custom computing machines for stream computation on FPGAs.

Acknowledgements This research was partially supported by Grant-in-Aid for Scientific Research (B) No. 23300012 and Grant-in-Aid for Challenging Exploratory Research No. 23650021 from the Ministry of Education, Culture, Sports, Science and Technology, Japan. We thank the support by ALTERA university program.

References

1. Takizawa, H., Yamada, N., Sakai, S., Kobayashi, H.: Radiative heat transfer simulation using programmable graphics hardware. In: Proceedings of the 5th IEEE/ACIS International Conference on Computer and Information Science, pp. 29–36 (2006)
2. TOP500. http://www.top500.org (2014)
3. Makino, J.: The GRAPE project. Comput. Sci. Eng. **8**(1), 30–40 (2006)
4. Shaw David E., et al. Anton, a special-purpose machine for molecular dynamics simulation. Commun. ACM **51**(7), 91–97 (2008)
5. Shirazi, N., Walters, A., Athanas, P.: Quantitative analysis of floating point arithmetic on FPGA based custom computing machines. In: Proceedings of the IEEE Symposium on FPGA's for Custom Computing Machines, pp. 155–162 (1995)
6. deLorimier, M., DeHon, A.: Floating-point sparse matrix-vector multiply for FPGAs. In: Proceedings of the International Symposium on Field-Programmable Gate Arrays, pp. 75–85 (2005)
7. Zhuo, L., Prasanna, V.K.: Sparse matrix-vector multiplication on FPGAs. In: Proceedings of the International Symposium on Field-Programmable Gate Arrays, pp. 63–74 (2005)
8. Dou, Y., Vassiliadis, S., Kuzmanov, G.K., Gaydadjiev, G.N.: 64-bit floating-point FPGA matrix multiplication. In: Proceedings of the International Symposium on Field-Programmable Gate

Arrays, pp. 86–95 (2005)
9. Sano, K., Pell, O., Luk, W., Yamamoto, S.: FPGA-based streaming computation for lattice boltzmann method. In: Proceedings of the International Conference on Field-Programmable Technology, pp. 233–236 (2007)
10. Murtaza, S., Hoekstra, A.G., Sloot P.M.: Compute bound and I/O bound cellular automata simulations on FPGA logic. ACM Trans. Reconfigurable Technol. Syst. **1**(4) (2009) Article 23.
11. Sano, K., Luzhou, W., Hatsuda, Y., Iizuka, T., Yamamoto, S.: FPGA-array with bandwidth-reduction mechanism for scalable and power-efficient numerical simulations based on finite difference methods. ACM Trans. Reconfigurable Technol. Syst. **3**(4) (2010)
12. Murtaza, S., Hoekstra, A.G., Sloot, P.M.: Cellular automata simulation on a FPGA cluster. Int. J. High Perform. Comput. Appl. **25**(2), 193–204 (2011)
13. Kono, Y., Sano, K., Yamamoto, S.: Scalability analysis of tightly-coupled FPGA-cluster for lattice Boltzmann computation. In: Proceedings of the 22nd International Conference on Field-Programmable Logic and Applications, August 2012. Paper W3B1.
14. Sano, K., Hatsuda, Y., Yamamoto, S.: Multi-FPGA accelerator for scalable stencil computation with constant memory-bandwidth. IEEE Trans. Parallel Distrib. Syst. **25**(3), 695–705 (2014)
15. Skordos, P.A.: Initial and boundary conditions for the lattice Boltzmann method. Phys. Rev. E **48**(6), 4823–4842 (1993)
16. Sano, K., Nishikawa, T., Aoki, T., Yamamoto, S.: Evaluating power and energy consumption of FPGA-based custom computing machines for scientific floating-point computation. In: Proceedings of the International Conference on Field-Programmable Technology, pp. 301–304 (2008)
17. Terasic Technologies. http://www.terasic.com (2013)
18. Altera Corporation. http://www.altera.com/literature/ (2013)
19. Inamuro, T., Yoshino, M., Ogino, F.: A non-slip boundary condition for lattice Boltzmann simulations. Phys. Fluids **7**(12), 2928–2930 (1995)
20. Noble, D.R., Chen, S., Georgiadis, J.G., Buckius, R.O.: A consistent hydrodynamic boundary condition for the lattice Boltzmann method. Phys. Fluids **7**(1), 203–209 (1995)
21. Chen, S., Wang, Z., Shan, X., Doolen, G.D.: Lattice Boltzmann computational fluid dynamics in three dimensions. J. Stat. Phys. **68**(3–4), 379–400 (1992)
22. Kobori, T., Maruyama, T., Hoshino, T.: A celluer automata system with FPGA. In: Proceedings of the IEEE Symposium on Field-Programmable Custom Computing Machines, pp. 120–129 (2001)
23. Sano, K., Kono, Y., Suzuki, H., Chiba, R., Ito, R., Koizumi, K., Yamamoto, S.: Efficient custom computing of fully-streamed lattice Boltzmann method on tightly-coupled FPGA cluster. ACM SIGARCH Computer Architecture News, **41**(5), 47–52 (2013)

Application of HPC to Earthquake Hazard and Disaster Estimation

Muneo Hori, Tsuyoshi Ichimura, Maddegedara L.L. Wijerathne, and Kouhei Fujita

Abstract This paper presents the application of high performance computing to the earthquake disaster and hazard estimation, which is the first step taken to mitigate or reduce earthquake disasters. Capability computing is used in the seismic response analysis of important structures. A non-linear dynamic finite element method is developed to solve a high fidelity model of such structures. Explained are two examples of such capability computing. Capacity computing is used as new method of earthquake hazard and disaster evaluation for a densely populated urban area. Integrated earthquake simulation, which seamless analyzes the wave propagation pro- cess, the structure seismic response process, and the reaction process for earthquake disaster. Two examples of such integrated simulations are presented.

1 Introduction

Development of new technology for the reduction and mitigation of earthquake disaster has been a task which is assigned to the earthquake engineering community. Even though relatively young, this community has made innovative contributions, such as aseismic design methodology, a base isolation system, and a crisis control/-management system. There are increasing demands for higher level of safety against earthquake disaster; for instance, non-malfunctioning of all building facilities or quick recovery, provided that a building is free from damage.

M. Hori (✉)
Earthquake Research Institute, The University of Tokyo, Yayoi 1-1-1, Bunkyo, Tokyo 113-0032, Japan

Advanced Institute of Computational Sciences, RIKEN, Minamijima Machi 7-1-26, Chuoku, Kobe City, Hyogo 650-0047, Japan
e-mail: hori@eri.u-tokyo.ac.jp

T. Ichimura • M.L.L. Wijerathne
Earthquake Research Institute, The University of Tokyo, Tokyo, Japan

K.Fujita
Advanced Institute for Computational Science, RIKEN, Kobe, Japan

M.M. Resch et al. (eds.), *Sustained Simulation Performance 2014*,
DOI 10.1007/978-3-319-10626-7_17

To meet demands for higher level of safety against earthquake disaster, the first step that has to be taken is to make more reliable estimation of possible earthquake hazard and disaster. Here, the hazard is seismic ground motion created by earthquake waves, and disaster is structure damage and malfunctioning induced by the seismic ground motion. It should be emphasized that while the earthquake hazard is caused by physical processes of the propagation of earthquake waves, the earthquake disaster involves both physical and social processes, from which direct and in-direct losses are induced, respectively.

At this moment, the earthquake hazard and disaster estimation uses empirical methods, which are based on statistical analysis of past data of ground motion and structure damage [3, 12]. Social and economic damage is evaluated in a similar manner. The reliability of the estimation made by these methods is limited. Indeed, seismic ground motion is categorized as a certain index, fully ignoring its physical characteristics, such as the duration time and the dominant frequencies. It is necessary to develop a more reliable method of the earthquake hazard and disaster estimation.

High Performance Computing (HPC) provides us a possibility of developing such a more reliable method of the earthquake hazard and disaster estimation [2, 8]. As for an extremely important structure, HPC enables us to make seismic response analysis which uses an analysis model of highest fidelity, so that possible local or global damages are estimated in finest spatial resolution. HPC also enables us to carry out earthquake simulation for an entire urban area, by analyzing hundred thousands models of structures which are located in the area for given multiple scenarios of a possible earthquake. These two are regarded as capability computation and capacity computing of HPC, respectively.

In this paper, we present the current state of applying capability computing and capacity computing of HPC to the earthquake hazard and disaster estimation. In the next section, we briefly summarize the background of these two computings, presenting governing equations, numerical methods, and model settings. In Sect. 3, we present two examples of capability computing, the seismic structure analysis of a nuclear power plant which solves a fault-structure system, and a reinforced concrete pier in which local cracking is analyzed. In Sect. 4, we present capacity computing of the urban area seismic response analysis, which uses various data source and analysis methods for earthquake hazard and disaster estimation. Concluding remarks are made in the last section.

2 Overview of HPC Application

2.1 Capability Computing

Structure seismic response analysis evaluates deformation and stress of a structure when ii is subjected to seismic ground motion. The mathematical problem of this analysis is described as a wave equation of solid continuum, which is given as partial differential equations. The simplest form of this equation is for a linearly elastic body, i.e.,

$$\rho \ddot{\mathbf{u}} - \nabla \cdot (\mathbf{c} : \nabla \mathbf{u}) = \mathbf{0}, \tag{1}$$

where $\mathbf{u}$ is a vector-valued displacement function, dot and ∇ stand for the temporal and spatial differentiation, respectively, and ρ and $\mathbf{c}$ are density and forth-order elasticity tensor; this equation becomes non-linear if $\mathbf{c}$ changes depending on the state of stress or deformation. Interaction between the structure and the on-site ground must be taken into consideration, and a model used for this interaction problem ought to consist of the structure and the surface ground layers. Stress evaluation is essential for local and global failure, and highest fidelity is required for the model, since the degree of stress concentration depends on the local configuration of the structure.

In view of the above mentioned characteristics, the structure seismic response analysis generally employs finite element method (FEM) as a numerical analysis tool, using a large scale model for a structure and surface ground layers; the degree-of-freedom of the model exceeds 10,000,000, even though it is in the order of 10,000 or 100,000 for practical cases of aseismic design. Hence, this analysis is categorized as capability computing. It should be mentioned that non-linearity of concrete or soil which are used for buildings and infrastructures is different from that of metallic materials which are commonly used for automobiles, ships, and airplanes. These materials exhibit a negative slop for the relation between deformation and stress, which often causes difficulties in carrying out numerical analysis.

The use of FEM is inevitable for the accurate evaluation of stress concentration in the structure seismic response analysis [9, 10, 13, 14]. Unstructured mesh is used for FEM, and it is a bottleneck of maintain the scalability of parallel computation. A major part of FEM computation is spent in *solver*, i.e., solving a matrix equation which is derived by discretizing the wave equation. Numerous researches have been made to develop an efficient solver in the research community of solid computational mechanics and applied mathematics.

2.2 Capacity Computing

Urban are seismic response analysis is a new method of evaluating earthquake hazard and disaster for a densely populated area [6, 7]. Unlike the structure seismic response analysis, this analysis has a nature of capacity computing, since (1) it must consider 1,000 or more scenario of a possible earthquake and (2) the number of buildings and structures which needs to be analyzed is in the order of 1,000,000. More numerical computation is needed if a more sophisticated analysis model is used for the buildings and structures, provided that we can construct such models using available data of the target area.

An earthquake scenario refers to failure mechanisms or focal mechanism at a target fault, from which earthquake waves are emitted. It is not possible to select one scenario that is going to happen. Therefore, there remain large amount of uncertainties in a possible earthquake, and we have to consider multiple scenarios. Building models have uncertainties as well, since the quality and quantity of the urban area data with which these models are constructed; as an extreme case, old buildings do not have digital data of their design. Even if data used in design of a building or a structure are available, there are some discrepancy between the designed values and the actual values of structure properties; for instance, strength is usually underestimated in design, to make extra safety. These are the sources of uncertainties in the building models. Due to the uncertainties of an earthquake and building models, it is necessary to make the urban area seismic response analysis for multiple scenarios of a possible earthquake and to use a set of analysis models for a building or a structure.

It should be mentioned that a specific seismic response analysis method has to be used for a particular class of buildings and structures (i.e., high or low buildings, railway or road bridges, or water or sewage pipeline networks). This is because each class of buildings and structure has particular characteristics of the seismic responses. While FEM is applicable to any kind of buildings and structures, it often uses a large scale model, the construction of which needs detailed data of its mechanical and structure properties. As an alternative, used is a specific numerical analysis method which has been developed for a certain class of buildings and structures and does not require detailed data of their properties; the earthquake engineering community reserves a set of such numerical analysis methods, which could be understood as asset of numerical analysis.

3 Structure Seismic Response Analysis

This section presents the following two examples of the structure seismic response analysis that is regarded as capability computing of HPC: (1) a fault-structure system of a nuclear power plant; and (2) a reinforced concrete pier for road bridges. The first example uses a large scale model for linear analysis, and the second example is for non-linear analysis that accounts for local failure due to cracking.

3.1 Fault-Structure System of Nuclear Power Plant

A fault-structure system consists of a target fault and a target structure; see Fig. 1 for a schematic view of the system [9, 10, 13, 14]. Analyzed are the whole processes of this system, namely, the fault rupture process, the wave propagation through the crust, the wave amplification near ground surfaces, and the resulting structure response, which is influenced by interaction of on-site surface layers. Macro Micro

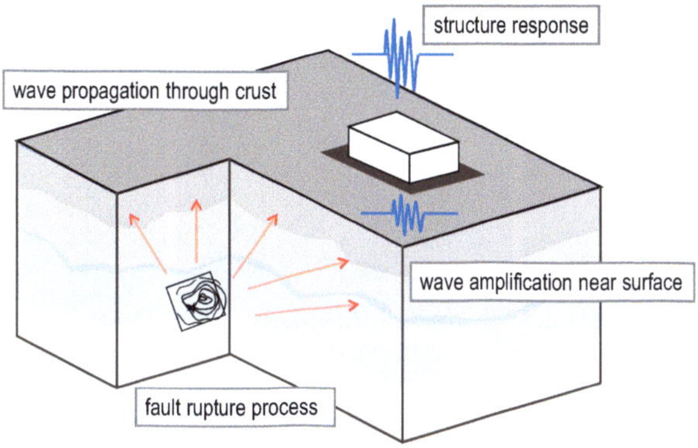

Fig. 1 Schematic view of analysis model of nuclear power plant

Table 1 Properties of analysis model

	Quantity	Memory usage
(a) Macro-analysis model		
Tetrahedral element	225, 361, 034	
Hexahedra element	96, 805, 249	101.8 Gb
Degrees of freedom	421, 940, 238	
(b) Micro-analysis model		
Tetrahedral element	2, 288, 210	
Hexahedra element	2, 138	29.8 Gb
Degrees of freedom	1, 455, 081	

Analysis (MMA), a large scale dynamic FEM program, has been developed to analyze the fault-structure system. This method is based on a multi-scale or two-scale analysis, as it first solves the entire system in a low resolution (macro-analysis) and then refines the solution for a structure in a high resolution (micro-analysis).

We study seismic response of an actual nuclear power plant, which is hit by 2007 Nigata-ken Chuetsu Oki Earthquake, applying MMA to an analysis model of a fault-structure system of this plant. The target domain is of the order of 100,000 m, while the size of elements used in the micro-analysis is of the order of 1 m; see Table 1 for the characteristics of the macro- and micro-analysis models. The model used for the macro- and micro-analysis is validated by comparing the observed records of an actual earthquake, while the FEM solver that is commonly used for MMA is verified by checking the accuracy of reproducing the analytical solution.

In Fig. 2, the seismic ground motion computed by the macro-analysis is presented; snapshot of the displacement norm distribution are presented from 7 to 21 s after the earthquake occurrence. It is clearly seen that the topographical effects change the pattern of the ground motion distribution, and relatively large ground motion hits the nuclear power plant at 7.0 and 10.0 s. The results of the micro-analysis are presented in Fig. 3; snapshots of the plant building in the EW and NS directions are presented in (a) and (b), respectively, from 5 to 9 s after the earthquake

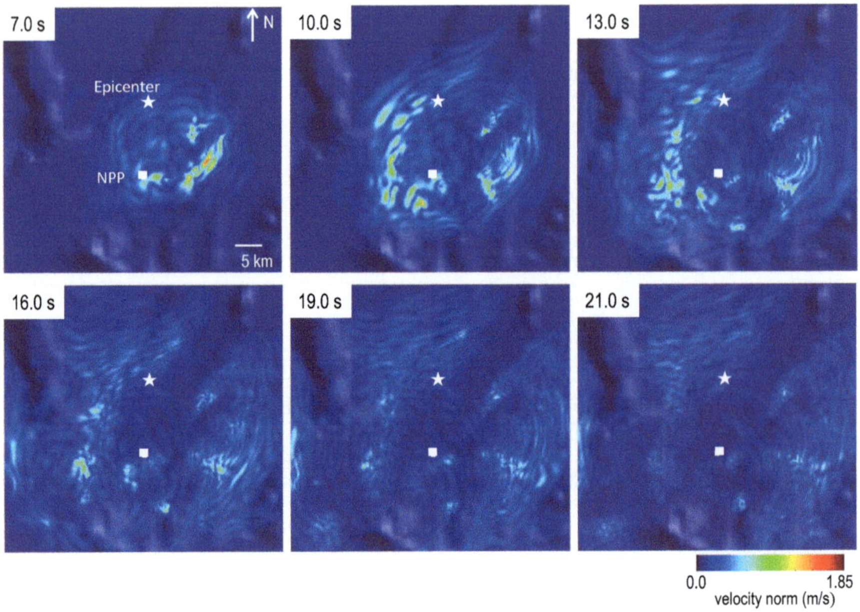

Fig. 2 Snapshot of displacement norm distribution computed by macro-analysis

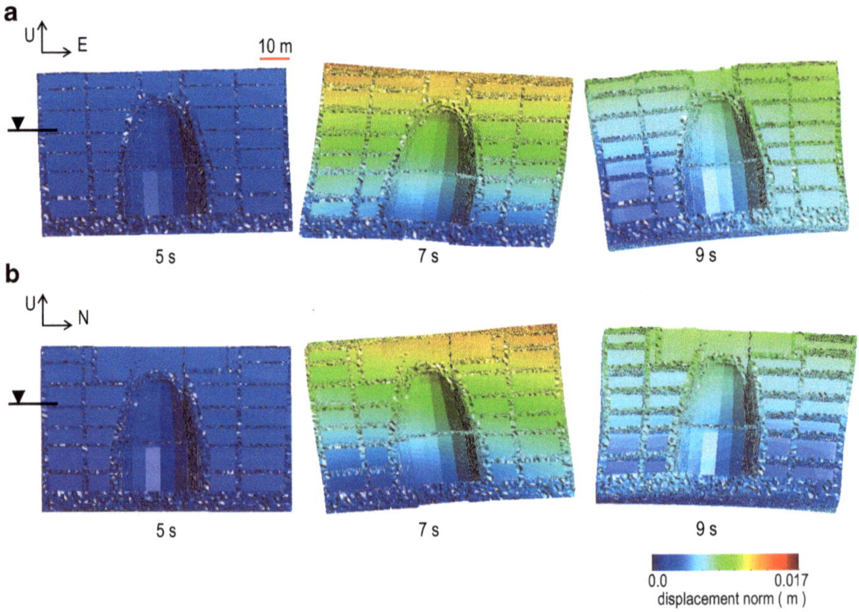

Fig. 3 Snapshot of power plant building deformation computed by micro-analysis. (**a**) EW direction, (**b**) NS direction

occurrence, and the building deformation is amplified by 1,000 times. As is seen, the building deforms more or less uniformly, and no concentration of deformation is found in these snapshots. This suggests that the integrity of the nuclear power plant building is maintained.

High cut filter of 1 Hz is applied to the results shown in the above, and hence there remain a large room to improve the temporal resolution, say, up to 10 Hz, in which the first natural frequency of the nuclear power plant building is included. Increasing the temporal resolution inevitably accompanies the increase in the spatial resolution due to the nature of constant wave velocity (i.e., wave length of higher frequency becomes shorter). We thus need an analysis model of finer discretization, which further requires larger scale numerical computation, in order to investigate the seismic resistance of this building more substantially.

3.2 Reinforced Concrete Pier

A bridge structure is generally divided into a deck (or an upper structure) and a pier (a lower structure) which includes foundations consisting of supports and piles. The pier is a core part for the seismic response analysis, as it carries large weight of the upper structure. Reinforced concrete (RC) is used for the pier. RC is composite of brittle concrete and ductile steel, and hence it exhibits good mechanical properties, even though concrete suffer from local cracking.

Analysis of cracking has been a tough problem in the community of solid computational mechanics. In particular, cracking in concrete results in complicated surfaces due to kinking and brunching of the propagating crack. It is certainly true that accurate analysis is possible if configuration of the crack surfaces is given. However, determining complicated crack surface configuration is a different problem. Since crack is formulated as discontinuity of a displacement function, Particle Discretization Scheme (PDS), which allows discontinuities in a discretized function, is developed for this purpose [5, 15]. PDS–FEM, which implements PDS into FEM, is used for non-linear analysis of RC structures when cracking and resulting failures are studied.

A real scale experiment has been made for the seismic response of an RC bridge pier subjected to extremely large ground motion. The reproduction of this experiments is studied by applying PDS–FEM to the analysis model shown in Figs. 4 and 5. As is seen, reinforcing steel bars are modeled by using sold element models; see Table 2 for the model characteristics. It should be mentioned that an actual steel bar is deformed, i.e., notches are periodically put to increase the connectivity between concrete and steel.

In Table 3, the performance of the PDS–FEM program is summarized; K computer is used, in which 1 node has 8 computing cores in it and hybrid parallelization is implemented in the program. As is seen, the strong scaling is 79.6 %, which is fairly good number of this class of FEM even though only a equator of the whole nodes of K computer is used. Figure 6 presents the distribution of crack

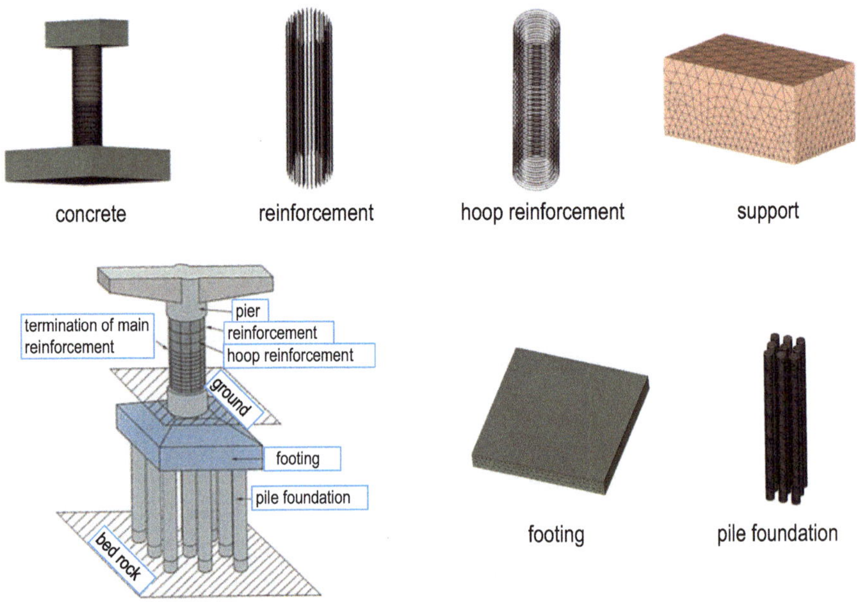

Fig. 4 Schematic view of analysis model of RC bridge pier

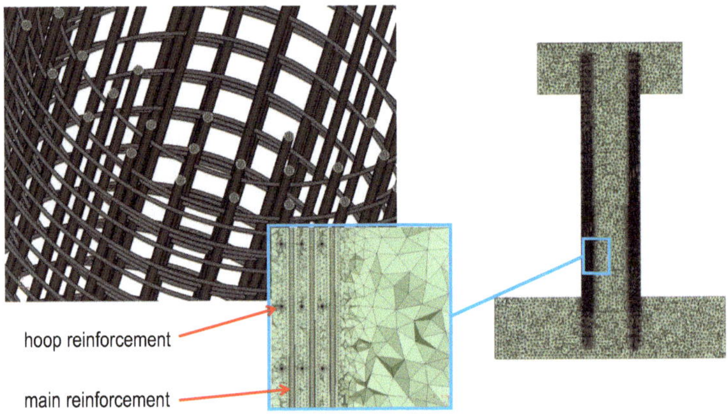

Fig. 5 Details of analysis model of RC bridge pier

surfaces when a relatively large ground motion is input. It is intuitively acceptable that cracking takes place in the lower part of the RC pier, since relatively large stress acts there. However, it is not acceptable that cracking occurs at the top of pile foundations. Interface condition between the support and the pile foundations must be improved to avoid such unrealistic cracking.

The results shown in the above must be regarded as tentative, since numerical treatment of the structure component connection (i.e., the support and the pile

Table 2 Properties of analysis model of reinforced concrete bridge pier

Member	Node	Element	Degree of freedom
Concrete			
Reinforcement	650,745	1,913,132	1,952,235
Hoop reinforcement	882,508	2,885,185	2,647,524
Support	6,159	30,242	18,477
Footing	44,796	252,664	134,388
Pile foundation			
Soil	1,027,196	5,751,000	3,081,588
Total	7,419,125	35,706,880	22,257,375

Table 3 Performance of PDS-FEM program in K computer

Node	Run time [s]	Performance [TFLOPS]	Peak performance ratio [%]	Scaling (strong)
12,288	204.5	89.7	5.71	–
24,576	128.3	151.3	4.81	79.6

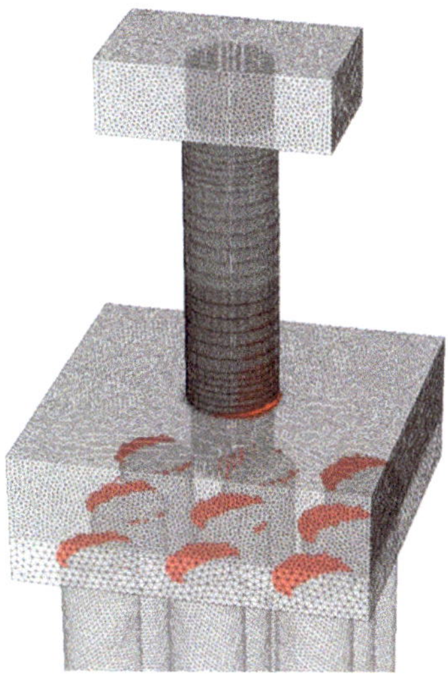

Fig. 6 Distribution of cracking in concrete bridge pier

foundation) must be improvement. Also, we have to study the effects of the reinforcement configuration on the bond between concrete and steel; a reinforcing bar is usually called a deformed bar as its surface is patterned. Constructing a high fidelity model of such a deformed bar results in drastic increase in the spatial discretization since the radius of the deformed bar is a few cm while the dimension of the pattern is a few mm. There surely exists an alternative of treating such details of an RC structure, but the use of an analysis model of larger scale is a solution. The analysis of such a model results in larger amount of numerical computation.

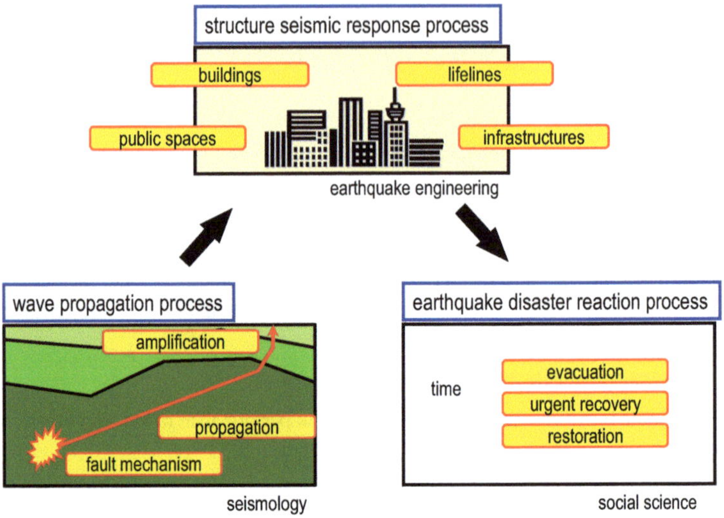

Fig. 7 Schematic view of IES

4 Urban Area Seismic Response Analysis

Urban area seismic response analysis is realized in the form of Integrated Earthquake Simulation (IES) [4, 6, 7, 11, 16], which seamlessly simulates earthquake hazard and disaster in an urban area for a given scenario of a possible earthquake; see Fig. 7. Computed are the following three phases of the earthquake hazard and disaster: (1) the wave propagation process; (2) the structure seismic response process; and (3) the reaction process for earthquake disaster. Suitable models of a target urban area are constructed for the numerical simulation of each process, by using available data which are often stored in form of Geographical Information System (GIS). IES takes full advantage of capacity computing, in order to carry out the seamless simulation for multi scenarios of a possible earthquake and multi analysis models.

4.1 Overview of Urban Area Seismic Response Analysis

Simulation of the earthquake wave propagation process uses solid element FEM and analyzes an underground structure model. Just like MMA, multi-scale analysis is made if the wave amplification in surface ground layers is computed up to a few Hz; non-linear material properties of soil are considered in the computation. A model of the surface layers is constructed for a given set of ground layer data. Some approximations are made in inter- or extrapolating the date, and the spatial resolution of the model reaches less than 1 m. Distribution of seismic ground motion, which is given as time series of three-dimensional displacement components, is accurately

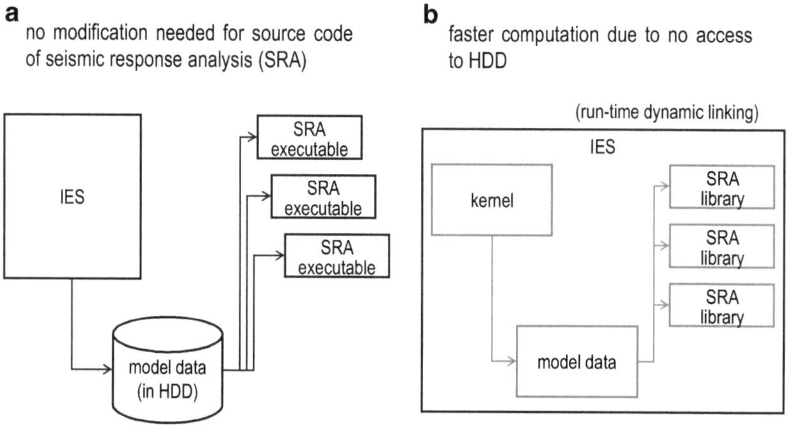

Fig. 8 Implementation of third party program into IES. (**a**) system call, (**b**) dynamic/static library

computed, so that topographical effects of the underground structures on ground motion are accounted for.

In simulation of the structure seismic response process, a suitable model is constructed for each building or structure in a target area, and it is analyzed by using a proper non-linear seismic response analysis. Synthesized ground motion of the earthquake wave propagation process is used as input ground motion. Depending on the quality and quantity of the available GIS in which building and structure data are stored, a model of suitable complexity is chosen; certain mechanical and structure properties must be guessed if relevant information is not available. Detailed assessment of earthquake disaster is made by considering the seismic responses of all the models, which includes local or global damage. It should be mentioned that the present IES uses system call or dynamic/static library to implement a third party program of the seismic response analysis for a particular class of buildings or structures; see Fig. 8.

The present IES implements two analysis methods for the reactions taken for earthquake disaster, namely, mass evacuation and recovery process of lifeline. Multi agent simulation (MAS) is employed [1]. In MAS, agents, which spontaneously act with distinct attribute and functionalities, are used as a model of an evacuee or an engineer, and a model for road network or lifeline pipeline network is constructed as an environment of MAS. The results of the structure seismic response process are taken into consideration, in constructing the environment; some parts of the road network is damaged or damages are assumed for certain segments of the pipeline network. MAS for the mass evacuation process could reveal bottlenecks in evacuation routes and evaluate effects of countermeasures for them, and MAS for the recovery process is aimed at considering a better strategy of recovery, by considering coordinated repair of different lifeline systems.

The urban area seismic response analysis is regarded as capacity computing when it carries out simulations for numerous scenarios of a possible earthquake.

In particular, the structure seismic response process is embarrassingly parallel computing since each analysis model is computed independently. Management of I/O is a key issue when it analyzes a larger number of building and structure models; for instance, the number of non-linear structure seismic analysis becomes 10^{12} if the number of analysis model for one building or structure is 10^3 (in order to account for its uncertainties), the number of buildings and structures in a target area is 10^6, and the number of the earthquake scenario is 10^3.

4.2 *Partial Reproduction of 2011 Great East Japan Earthquake Disaster*

As an example of integrated simulation, we carry out numerical simulation in order to partially reproduce 2011 Great East Japan Earthquake Disaster. The main objective of this simulation is to validate the present IES, by comparing the simulation results with the observed data; the comparison is being made at this moment. Thus, seismic response analysis and tsunami inundation analysis in an urban area are integrated in this example.

Figure 9 provides a schematic view of an analysis model used for the two analyses. The seismic response analysis employs a third party program of non-

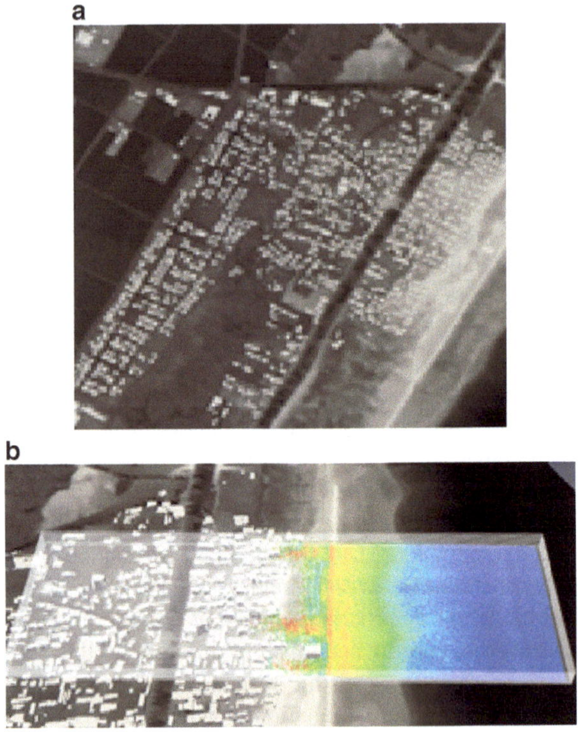

Fig. 9 Schematic view of urban area models used for partial reproduction of 2011 Great East Japan Earthquake Disaster. (**a**) Urban model for seismic response analysis. (**b**) Urban area model for tsunami inundation analysis

linear multi-degree-of-freedom system (which has been developed by professional structure researchers), and a suitable analysis model is constructed for each houses; while GIS data are available for the configuration of the house, no information about the structure properties is open, and parameters needed for the third party program are made (with the help of the professional structure researchers). The tsunami inundation analysis employs Smoothed Particle Hydrodynamics, and the houses as well as the surface ground are modeled by using rigid body particles. It should be mentioned the model construction is automated; these two urban areas are made from the same GIS by applying different programs of the model construction.

Snapshots of the seismic response analysis and the tsunami inundation analysis are presented in Fig. 10; each house is shaken according to its own structure properties (that is mainly categorized as the natural frequency and the corresponding

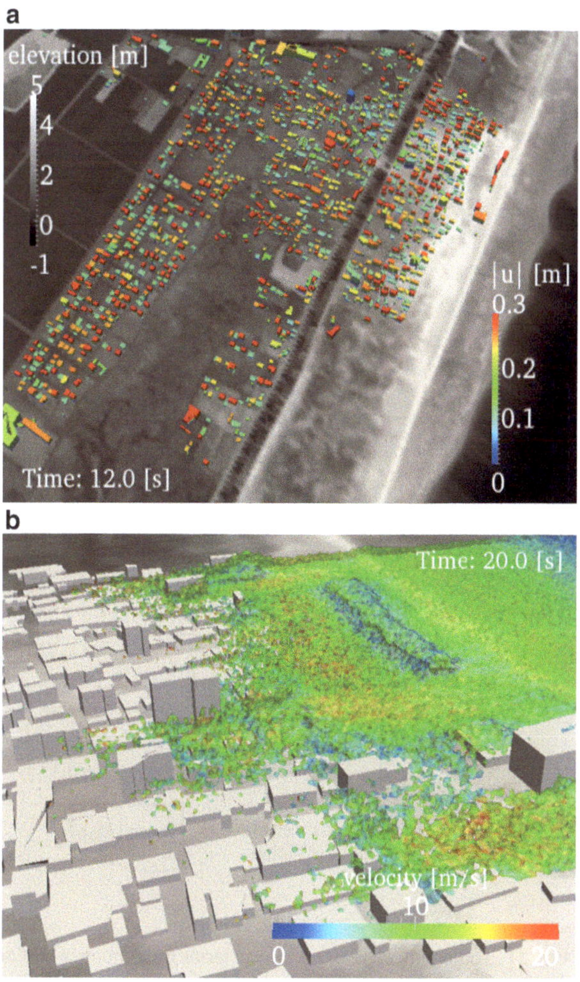

Fig. 10 Results of IES for partial reproduction of 2011 Great East Japan Earthquake Disaster. (**a**) Snapshot of seismic response analysis. (**b**) Snap shot of tsunami inundation analysis

mode) and the flow of the tsunami inundation is influenced by the existing houses. As mentioned, the comparison of these results with the observed data are on the process. At least, however, it is seen that the two urban area models are simulated by using IES. HPC application plays a key role in extending the size of the target area or increasing the number of an analysis model for each building in the seismic response analysis. Moreover, HPC enables us to make coupled simulation of the seismic response analysis and the tsunami inundation analysis, since the inundation flow is changed depending on the integrality of the houses shaken by the ground motion.

4.3 *Partial Estimation of Tokyo Metropolis Earthquake*

Figure 11 presents a schematic view of the seamless computation of IES which is being developed to carry out partial estimation of a possible Tokyo Metropolis Earthquake. It consists of the three analysis, namely, the ground motion, the seismic response and the mass evaluation, which are designated as an analysis module. The corresponding urban area model is indicated as an analysis data, which is constructed by an assembly of several data sources. It is the assembly that is in charge of automated model construction of the urban area models by using available data.

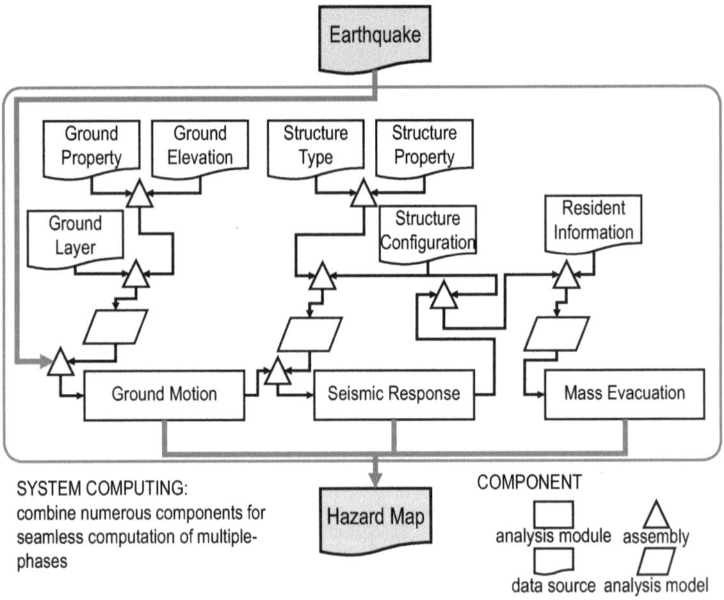

Fig. 11 IES architecture expressed in terms of components of analysis modules, analysis models, data sources, and assembly

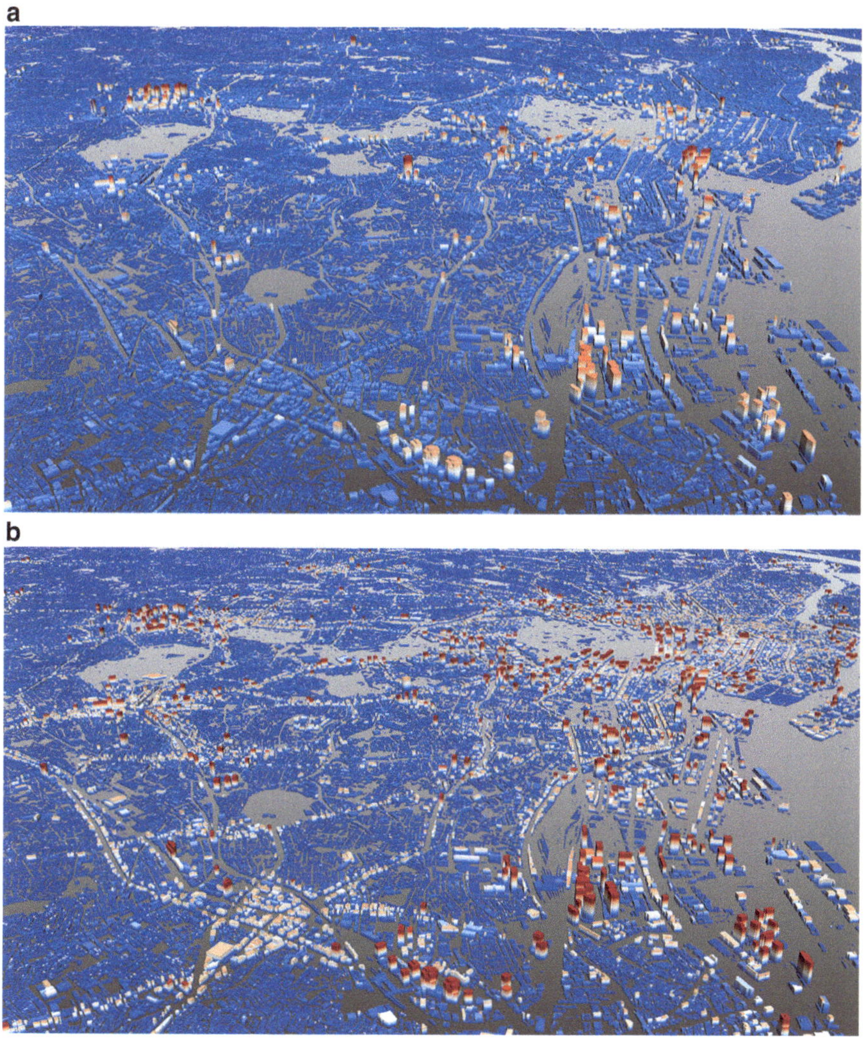

Fig. 12 Snapshots of urban area seismic response computed for Tokyo Metropolis Earthquake. (**a**) 1.25 s, (**b**) 1.45 s

As an example of the partial estimation, we apply this IES to Tokyo 23 Wards for the seismic response analysis. The number of the building models analyzed exceeds 1,000,000; see Fig. 12 for snapshots of the urban area seismic response analysis made for Tokyo 23 Wards. Like the urban area model shown in Fig. 9, third party programs of a linear/non-linear multi-degree-of-freedom system is employed, and the models are automatically constructed by using available GIS.

Taking advantage of being embarrassingly parallel, the seismic response analysis is carried out for each GIS tile of Tokyo Metropolis, the size of which is of the order

Table 4 Average runtime of IES for Tokyo Metropolis Earthquake

	Average run time (min)
Simulation of 1 GIS tile	10
Simulation of Tokyo	210
Visualization of Tokyo	60

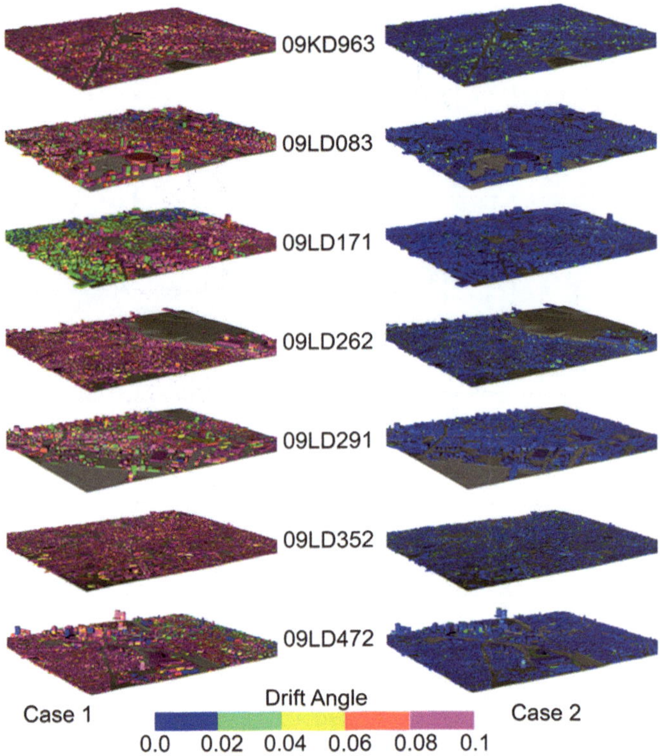

Fig. 13 Results of IES for partial estimation of Tokyo Metropolis Earthquake

of 1,000 m. In a moderately large scale parallel computing system (consisting of 64 nodes of SGI Altix UV 1000), the runtime of this analysis is summarized in Table 4; 1 node, which consists of 64 CPU's with six computing cores, is used for the computation of one tile. It is seen that the present urban area simulation is fast enough to make an urgent evaluation of earthquake disasters that is made right after the occurrence of a large earthquake.

A major concern of the urban area seismic response analysis is to examine how the structural damage changes depending on the ground distribution. Two examples of computing the damage distribution are presented in Fig. 13; a drift angle of the building, which is a standard index of the structural damage, is used. It should be emphasized that two seismic ground motions which are actually measured and have the peak ground acceleration of the same degree are used. These two ground motions are input to all the analysis models, in order to emphasize the difference in the structural damage distribution. As is seen, the damage distribution of all the GIS tiles is different from each other, even though seismic ground motion of similar intensity is used.

Conclusion
Although further studies are surely needed, the earthquake hazard and disaster evaluation that is made by the structure and urban area seismic response analysis are more scientifically rational, compared with the existing evaluation that uses empirical relations of the earthquake hazard and disaster. This is mainly because the spatial and temporal resolution of the evaluation increases as models of higher fidelity are used. The analysis of such analysis models surely need HPC.

In the earthquake engineering community, numerical computation is being regarded as a research tool. However, the utilization of HPC that strengthens this tool is limited, since the construction of an analysis model for structure or urban area seismic response analysis needs laborious works. Together with increasing the speed and scale of the numerical computation by making use of HPC, we need a sophisticated method of constructing analysis models from available data.

References

1. Aguilar, L.,E., Wijerathne, M.L.L., Hori, M., Ichimura, T., Tanaka, S.: On the development of an MAS based evacuation simulation system: autonomous navigation and collision avoidance. PRIMA 2013: Principles and Practice of Multi-Agent Systems, Lect. Notes Artif. Intell. **8291**, 388–395 (2013)
2. Fujita, K.: Enhancing integrated earthquake simulator for high resolution disaster estimation of large cities. Ph.D. Thesis, Department of Civil Engineering, The University of Tokyo (2014)
3. GEM Foundation, GEM. (2014) http://www.globalquakemodel.org/
4. Homma, S., Fujita, K., Ichimura, T., Hori, M., Citak, S., Hori, T.: A physics-based Monte Carlo earthquake disaster simulation accounting for uncertainty in building structure parameters, Proc. Comput. Sci. **29**, 855–865 (2014). doi:10.1016/j.procs.2014.05.077
5. Hori, M., Oguni, K., Sakaguchi, H.: Proposal of FEM implemented with particle discretization scheme for analysis of failure phenomena. J. Mech. Phys. Solids **53**, 681–703 (2005)
6. Hori, M., Ichimura, T.: Current state of integrated earthquake simulation for earthquake hazard and disaster. J. Seismol. (2008). doi:10.1007/s10950-007-9083-x
7. Hori, M.: Introduction to Computational Earthquake Engineering, 2nd edn. Imperial College Press, London (2011)
8. Hori, M., Wijerathne, M.L.L., Ichimura, T., Tanaka, S.: On multi agent simulation for estimation of human action against earthquake disasters. J. Earthquake Tsunami (2013). doi:10.1142/S1793431113500085
9. Ichimura, T., Hori, M., Bielak, J.: A hybrid multiresolution meshing technique for finite element three-dimensional earthquake ground motion modeling in Basins including topography. Geophys. J. Int. **177**, 1221–1232 (2009)

10. Ichimura, T., Fujita, K., Hori, M., Sakanoue, T., Hamanaka, R.: Three-dimensional Nonlinear Seismic Ground Response Analysis of Local Site Effects for Estimating Seismic Behavior of Buried Pipelines, J. Pressure Vessel Technol. **136**, 4 (2014). doi:10.1115/1.4026208
11. Sobhaninejad, G., Hori, M., Kabeyasawa, T.: Enhancing integrated earthquake simulation with high performance computing. Adv. Eng. Softw. **42**, 286–292 (2011)
12. The Federal Emergency Management Agency, Hazus. (2014) http://www.fema.gov/hazus
13. Quinay, P.E., Ichimura, T., Hori, M.: Waveform inversion for modeling three-dimensional crust structure with topographic effects. Bull. Seism. Soc. Am. **102**, 1018–1029 (2012)
14. Quinay, P.E., Ichimura, T., Hori, M., Wijerathne M.L.L., Nishida, A.: Seismic strucural response analysis considering fault-structure system – application to nuclear power plant structures. Prog. Nucl. Sci. Technol. **2**, 516–523 (2012)
15. Wijerathne, M.L.L., Oguni, K., Hori, M.: Numerical analysis of growing crack problem using particle discretization scheme. Int. J. Numer. Methods Eng. **80**, 46–73 (2009)
16. Wijerathne, M.L.L., Hori, M., Kabeyazawa, T., Ichimura, T.: Strengthening of parallel computation performance of integrated earthquake simulation. J. Comput. Civil Eng. **27**(5), 570–573 (2012)

Geometry Dependent Computational Study of Patient Specific Abdominal Aortic Aneurysm

Nisarg Patel and Uwe Küster

Abstract Computational approach for analyzing and modeling haemodynamic phenomena in an abdominal aortic aneurysm (AAA) is becoming more prevalent in today's medical industry and scientific community. AAA is a local dilatation of an aorta in abdominal region. Due to this degenerative disorder the capacity of aorta to withstand the mechanical forces decreases resulting into rupture at any time. There is a great demand in understanding the physiological cause, determining the factors which influence growth causing rupture and thus the treatment of AAA. There is no general consensus to determine the actual cause of aneurysm rupture in treated or untreated patients. However, vascular wall stress analysis and maximum diameter are considered two influential factors leading to rupture risk in field of computational study. Therefore it is of utmost importance to understand the behavior of blood flow in the aneurysmal aorta. The work presented here is the preliminary approach towards constructing a realistic patient specific model for fluid simulations. The approach concentrates on how the geometrical differences in vascular structure can lead to pathological differences in results which could ultimately change the wall stress dynamics. The fluid simulation of blood flow results show us how flow patterns and wall shear differ when considering different degree of complexity of AAA model, for example the aortic arc and thoracic aorta in the computational domain. And the study also shows the necessity of implementing noise filters on raw medical data to give us accurate segmented vascular structures. The results show significant differences in behavior of the flow patterns in geometrical models with varying complexity.

1 Introduction

Local irreversible expansion of an abdominal aorta results in a weak aortic wall which in turn decrease it's capacity to withstand the mechanical forces, thus causing rupture in the wall at any time. The rupture of an aorta is a life threatening scenario

N. Patel (✉) • U. Küster
High Performance Computing Center Stuttgart (HLRS), Nobelstrasse 19, 70569 Stuttgart, Germany
e-mail: patel@hlrs.de; kuester@hlrs.de

M.M. Resch et al. (eds.), *Sustained Simulation Performance 2014*,
DOI 10.1007/978-3-319-10626-7_18

where there is 80–90 % probability of casualties resulting into death of majority of patients before reaching the hospital [5]. Even if a patient receives medical aid in time, the percentage of deaths in such cases vary from 45 to 60 % [15]. When the diameter of an aorta increases gradually growing to 4.5 cm in diameter, treatment of aneurysms via endo-vascular surgery (EVAR) is generally recommended by medical institutes [1, 16]. But in many cases, it has been observed that the aorta ruptures even before reaching the 4.5 cm mark. Many researchers have shed light on the rupture of smaller aneurysms and in turn the prescription of EVAR [6]. Hence there is no general acceptance of a procedure or an ideal approach to determine risk factors causing the aneurysm rupture in a diagnosed patient. To determine these risks, a computational study can be helpful which facilitates better understanding of mechanobiology of flow behavior and wall mechanics.

Since 1980 researchers have implemented numerical and computational approaches in determining pathological phenomena in vascular structures [11–13]. Fluid structure simulation studies until now has been focused either on determining the tensile forces acting on the wall of aorta, finding the localized wall stress [10], predicting the influence of inlet angle of aneurysm or determining wall stress due to the diametric influence [7]. Few researchers have tried determining risk factors for aneurysm rupture by considering the geometrical parameters other than the diameter [2]. There are simulation results pertaining to use of wall stress calculated at the circumference of aneurysm and patient specific simulations predicting the stress distribution at local wall curvature [17]. It has also been presented that the peak wall stresses predict rupture risks more precisely than the diameter of an aneurysm sac. This signifies the fact that maximum diameter alone is not a sufficient criteria to determine rupture risk [3].

It is well known fact that the geometrical modeling plays an important role in the process of simulation and analysis of fluid structure simulation (FSI) or only fluid simulation. Yet, we have observed that majority of the researchers give little attention towards upstream geometric morphology, e.g. aortic arc and thoracic aorta. Significance of an upstream flow governed by the morphology of upstream aortic structure, starting from the heart to a dilated aorta, is neglected unintentionally to reduce the computational time and cost of the simulations. There has been a new trend in research which emphasizes on modeling realistic patient specific analysis and preserving geometric model characteristics like the thickness and the diameter of an aneurysm. There is little material available focusing on the significance of the upstream flow arising from the heart to aneurysm which induces differences in flow behavior in aneurysm.

This work includes the segmentation process for extraction of vascular geometry from the medical image acquired from computer tomography data as explained in Sect. 2. Flow dynamics along with FE-model is presented in Sect. 3. Pathological variations highlighting the differences in complex flow pattern and the tracing of the fluid particles in patient specific models are covered in Sect. 4. We have carried out simulations of two patient specific cases and presented results focusing on different degrees of complexity of the geometric models. Differences in complexity of geometric models are the differences in extending features of the aortic structure like aortic arc, thoracic aorta alongside to abdominal aorta.

2 Image Modeling

2.1 *Image Acquisition and Segmentation*

Two image data sets used in this work were acquired from Robert Bosch Hospital, Stuttgart, Germany. Computed tomographic (CT) images were used to extract the aortic structure as it provides good image contrast by using smoother convolution kernels (low image sharpness) during reconstruction of three dimensional (3D) data set. Both data sets were reconstructed using 512×512 resolution in transverse plane of human body. The increment in vertical direction of human body i.e. spacing between two image slices were kept 0.3 and 0.6 mm respectively.

2.2 *Image Processing*

Image processing is the central step for feature extraction, recognition and ultimately to characterize a precise vascular geometry from CT data. The acquired 3D data set is not sufficient to extract the computational geometry of an aortic structure as it comes with undesired artifacts (noise) which can hamper interpretation of data sets leading to high numerical and computational labor. Presented below are two filtering techniques implemented by us to achieve a smooth geometry of the desired aortic structure.

2.2.1 Filtering Techniques

There are filtering techniques like low pass filters, high pass filters, linear filters and various other filters available for enhancing the data set without excessive alteration in the original data set. But there is no single filtering technique which fulfills our requirements of extracting the vascular structure. Extracting vascular structure implies generating a Finite element (FE) model from the extracted threshold of an image data set. Hence we have concatenated two filters on the image data set to get smoother images and reduced noise, thus achieving smoother FE models without any artifacts.

Median Filter

The median filter is a low pass filter which effectively reduces noise and preserves edges. It is applied by finding the median in the prescribed patch window of different sizes and sliding the patch window on entire data set. It can be expressed as

$$I(P_{i,j,k}) = median\left\{P_{I,J,K} \mid (I,J,K) \in patch_{i,j,k}\right\} \tag{1}$$

Where, $patch_{i,j,k}$ is the number of pixel (for 2D image) or voxels (for 3D image) in the patch. Patch sizes in a two dimensional (2D) image are shown in Fig. 1. Filter is applied on the data set with different patch sizes which successfully preserves the topology of all blood vessels. Patch is a section formed by a center pixel/voxel and its neighborhood pixels/voxel which surrounds this centered pixel/voxel. Figure 1 depicts a 7 × 7 grid patch on left side and a 5 × 5 grid patch on the right side.

The size of blood vessels in abdominal and thoracic region are large compared to iliac vessels, hence we can apply filters with varying patch sizes. The application of patch size depends roughly upon the size of the vascular vessel which needs to be extracted. Therefore we can distinguish blood vessels in three different categories according to the size; tiny vessels, large vessels and aneurysm. Figures 2 and 3, demonstrates the effects of different patch sizes and its resulting output as a 2D

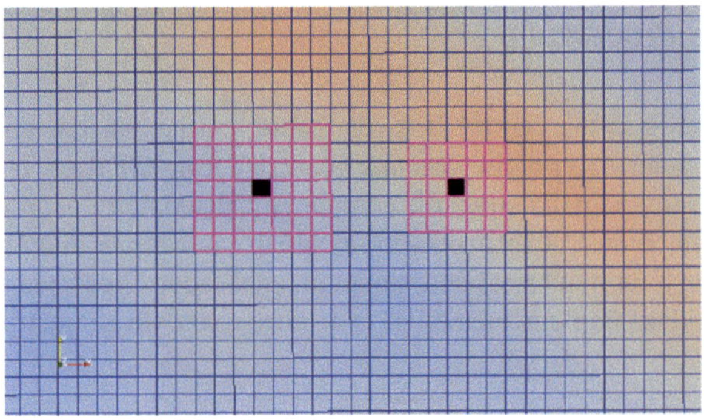

Fig. 1 2D image of sizes of patch (7 and 5) and center pixel in a patch

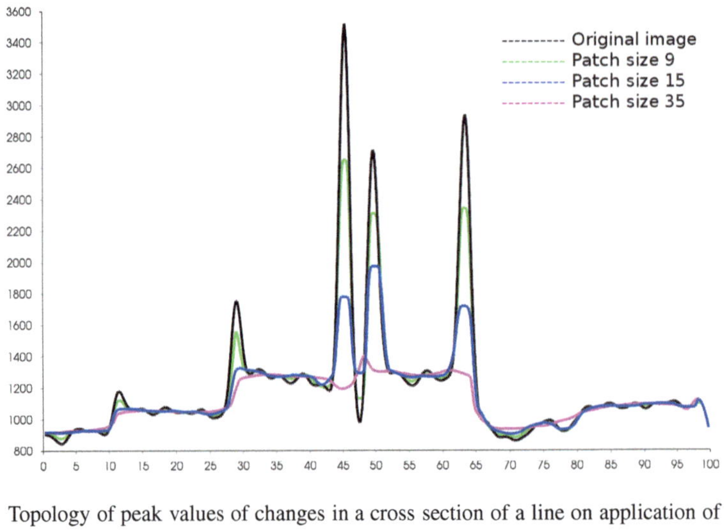

Fig. 2 Topology of peak values of changes in a cross section of a line on application of Median filter

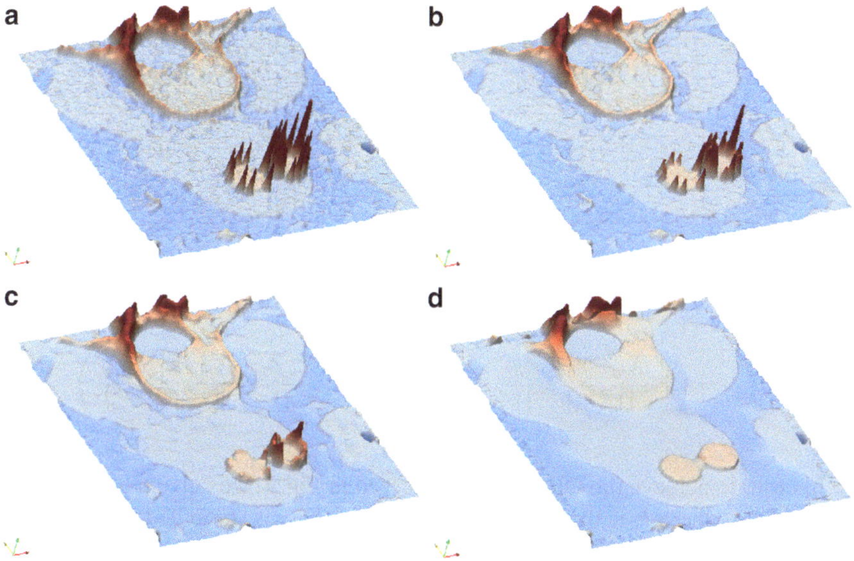

Fig. 3 Median filter applied on 2D scalar valued slice: **(a)** Original 2D image, **(b)** image with patch size 9, **(c)** image with patch size 15, **(d)** image with patch size 35

image. Figure 2 shows the removal of extremely high and low intensity pixels/voxel from image data set due to implementing this filter. Application of such median filter is demonstrated in Fig. 5b on a 3D data image.

QR Filter

The QR filter is a localized filter. It is based on obtaining the result image by solving the system of equations as shown in Eqs. (4)–(7). Where the A matrix is a Vandermonde matrix formed by a sequence of general second order polynomial for each point in convolution patch as can be seen in Eq. (2). $A^{\dagger}$ is a Moore-Penrose pseudoinverse of a Vandermonde matrix and the pseudoinverse $A^{\dagger}$ is given by upper triangulated matrix R^{-1} and orthogonal matrix Q^T via QR factorization. To filter the noise, polynomials of order 2 are taken into account and a Vandermonde matrix is formed for the number of voxels in the patch as defined in Eq. (3). The purpose of applying the QR filter in succession to median filter is to have localized smoothness of a structure, to distinguish the boundaries in detail.

The Vandermonde matrix can be built by all variables in Eq. (3) where p is the order of polynomial to implement, α is set of unknown coefficients and σ is the number of voxels in the patch.

$$A_{(m,:)} \times \left(\alpha_{(i,j,k)}\right)_{i,j,k} = \sum_{k=0}^{p} \sum_{j=0}^{k} \sum_{i=0}^{j} \left(\frac{X_m^i \times Y_m^j \times Z_m^k}{i! \times j! \times k!} * \alpha_{(i,j,k)} \right) \tag{2}$$

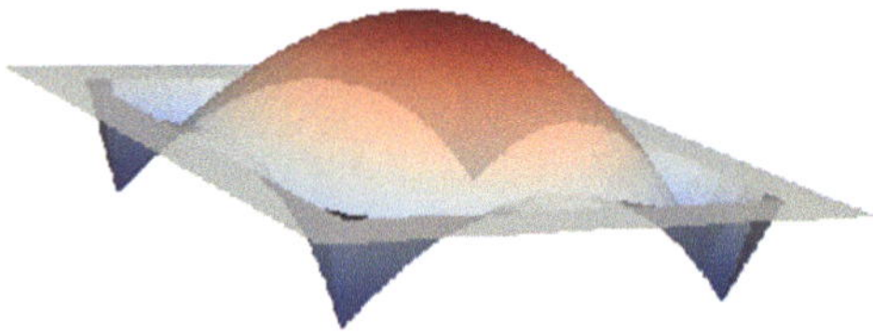

Fig. 4 QR patch convoluted on a 3D image data

Considering second order polynomial the variables of matrix are

$$A_{(m,:)_{m=1,\sigma}} = \left(1, \frac{X_m}{1!}, \frac{Y_m}{1!}, \frac{Z_m}{1!}, \frac{X_m Y_m}{1!}, \frac{X_m Z_m}{1!}, \frac{Y_m Z_m}{1!}, \frac{X_m^2}{2!}, \frac{Y_m^2}{2!}, \frac{Z_m^2}{2!}\right)_{m=1,\sigma} \tag{3}$$

In order to achieve the QR patch as observed in Fig. 4, we solve a linear system of equations where F is the image load vector and F_{new} is smoothed image vector,

$$A \times \alpha = F \tag{4}$$

$$\alpha = A^{\dagger} \times F \tag{5}$$

$$\alpha = (R^{-1} * Q^T) \times F \tag{6}$$

$$A_{(\frac{m+1}{2},:)} \times \alpha_{(:)} = \underbrace{A_{(\frac{m+1}{2},:)} \times (R^{-1} * Q^T)}_{\text{QR patch}} \times F_{(:)} = F_{new(:)} \tag{7}$$

In Eq. (7), $\frac{m+1}{2}$ denotes the center pixel/voxel in the patch. In a 3D data set, each voxel in the image is processed with corresponding neighboring voxels in the patch. This iterative processes of convolution effectively smooths the features. Figure 5, shows the application of QR filter concatenated with a median filter on AAA geometry. Convolution by using different patch sizes play a significant role in the outcome of smoothed images. Table 1, provides differences in the extracted volume of an aorta on application of the filters with different patch sizes. The volume of an entity does not vary more than 1 % on application of different sizes of patch. The percentage of volume change of a vessel when applying both the filters with different patch sizes is presented in Table 1. With these ratios of differences we can choose the size of the patch for tiny and large vessels respectively. For small vessels, change in volume is less then 0.01 % and for large vessels, patch size is selected for a change in volume less then 0.5 %. Volume shrinkage in smaller vessels affects the topology more than in bigger vessels. The QR filters were applied on high resolution 3D data sets by using patch size from 5 or 7 for smaller vessels and 9 or 11 for bigger vessels as can be seen in Fig. 5.

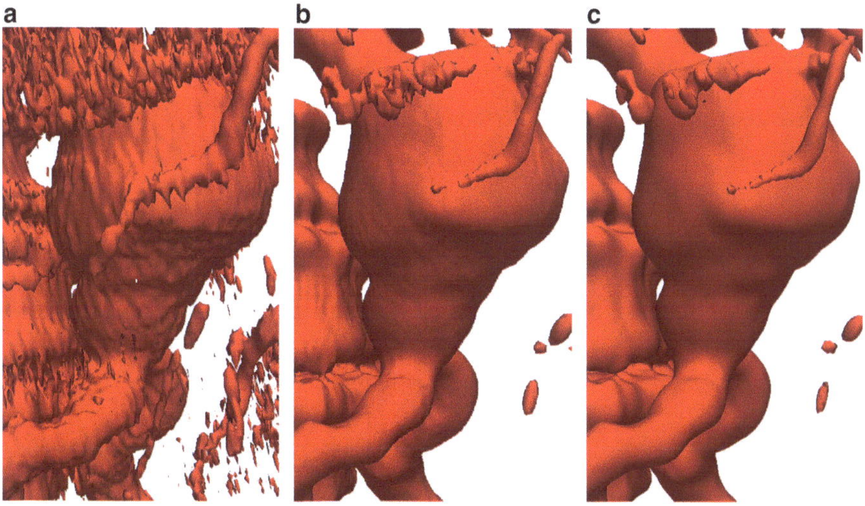

Fig. 5 Iso surface extraction from original data set showing processed image data set results: (**a**) original image, (**b**) median filtered, (**c**) QR filter linked after median

Table 1 QR and median filter with different patch size applied on a small image data set

Patch size for QR and median	Matrix size for each filter	Volume of geometry in mm^3	Difference in %
0[a]	–	37,532	0
5 and 5	125	37,549	0.045
7 and 7	343	37,567	0.093
9 and 9	729	37,601	0.183
15 and 15	3,375	37,756	0.596
19 and 19	6,859	37,843	0.828

[a]Original image without any filter

2.2.2 Computing Resources

From extensive simulations we have been able to conclude that QR algorithm is highly stable with the computational complexity of $O(n)$. It is important to note that the QR filter is formed by inverting the matrix once and thereafter an iterative process is applied. Dealing with high resolution data of size $512 \times 512 \times 1{,}024$ ($n = 268{,}435{,}456$ no. of voxels) is particularly time consuming. Iterative process for such large data set is carried out by using different machines with single core, the execution time for both filters is presented in Fig. 6b. The median filter is way more expensive as the quick sorting of neighboring pixels in the patch takes place for each pixel. The execution was carried out on four different machines as named; Intel Sandy Bridge with 32 GB mem (sb), Intel Sandy Bridge with 64 GB mem (mem64gb), Intel Nehalem 24 GB mem (mem24gb), Intel Nehalem 48 GB mem

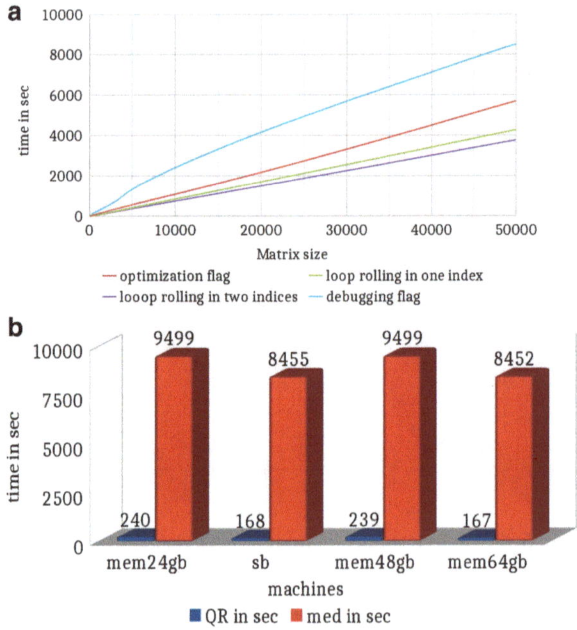

Fig. 6 Performance measurement and time consumption for single core process: (**a**) optimization for large matrices, (**b**) execution time

(mem48gb). Figure 6a shows the normal execution times and manually optimized (loop unrolled in two dimensions) time for filters of different matrix sizes. Figure 6b shows execution times for each filter by using a single core with optimization flag($-O3$) for the GNU FORTRAN compiler.

3 Computational Modeling

Patient specific geometry (PSG) development in computational modeling plays a crucial role in understanding the difference of AAA morphology results in the various computational analysis. Two PSG of AAA were extracted from CT images, termed AAA1 and AAA2, via semi automated iso-surface contouring using `Paraview`[1]. For each of the extracted AAA two finite element (FE) models, complex and simplified as shown in Fig. 7, were developed using the `ANSA`[2] pre-processor. The difference between both models is the inclusion of geometrical features such as aortic arc and thoracic aorta in the computational models. Inlet of both complex models (CM) are located at the beginning of the arc of aorta as shown in Fig. 7b, d. Inlets of both simplified models (SM) are restricted to the beginning

[1] http://www.paraview.org.

[2] http://www.beta-cae.gr/ansa.htm.

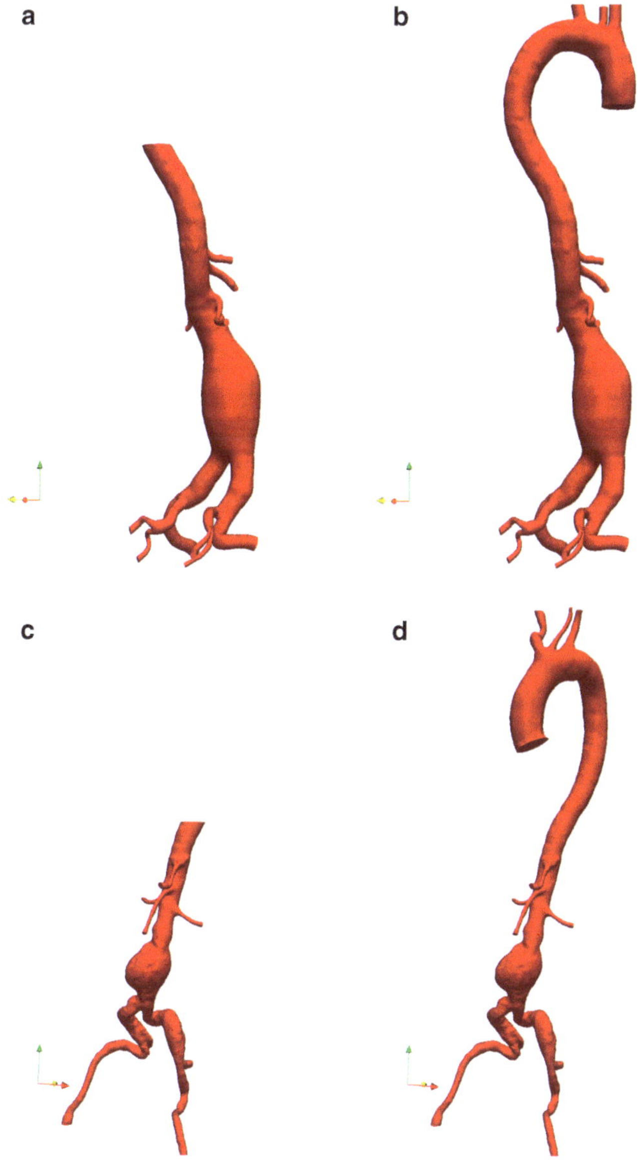

Fig. 7 Simplified and complex luman geometry of AAAs: (**a**) AAA1 simplified FE model without aortic arc, (**b**) AAA1 complex FE model with aortic arc, (**c**) AAA2 simplified FE model without aortic arc, (**d**) AAA2 complex FE model with aortic arc

of the abdominal aorta as can be seen in Fig. 7a, c. Outlets are located at the distal end of the main bifurcations of aorta. Modeling is restricted to luman geometries in order to determine the pathological differences in SM and CM.

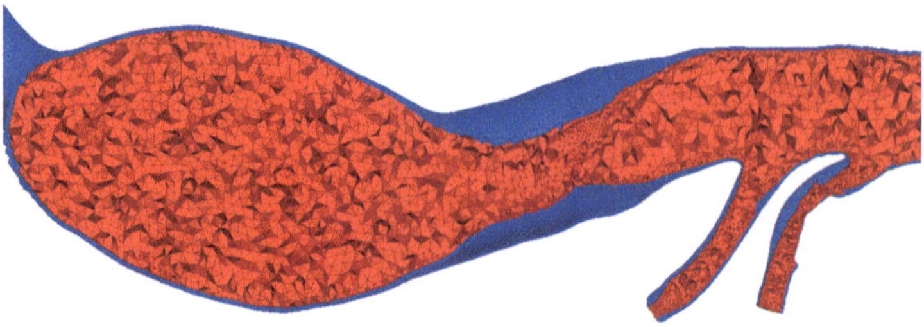

Fig. 8 FE mesh representation in a model

3.1 *Finite Element Model*

FE models with varying levels of complexity are shown in Fig. 7. The mesh topology of the luman domain were chosen such that the small triangulated and tetrahedral elements have been adopted at critical regions like the bifurcations, aneurysm and tiny arteries. A layered mesh was employed to capture the flow phenomena near the surface as shown in Fig. 8. A mesh dependency study was implemented on one of the AAA models before deciding the final mesh, see Table 2. Pressure and velocity distributions for four different mesh sizes and convergence criteria was compared. Pressure, velocity distribution and wall shear seems to remain stable in all the four mesh variations without any substantial differences. To capture the differences in fluid parameters, like turbulent mixing and circulation zones accurately in critical domain like aneurysm and near bifurcating regions, we choose varying mesh sizes with respect to AAA's geometry which are detailed in Table 3. FE models for SM and CM were consistent in mesh dimensions.

3.2 *Fluid Simulation Model*

3.2.1 Boundary Conditions

Theoretical pressure and velocity wave forms are implemented as boundary conditions which represent the blood flow as shown in Fig. 9. Outlet wave, by Olufsen [9], is applied at all terminals of large arteries. Inlet wave for the simplified models, refer Fig. 9b, c, are interpolated from respective CM inlet waveform. The pressure varies from 80 to 120 mmHg and velocity wave forms fulfill our aim of implementing the periodic wave forms to simulate the blood flow. Simulation is carried out for five cardiac cycles with one cycle each of 1.2 s. Fluid luman surface is modeled as a no-slip wall without structure of the vessel.

Table 2 AAA1 with different surface mesh sizes

Model	Max element size in mm	Min. element size in mm	mean element size in mm	Total number of FE element in million
AAA1 complex	6.5	0.25	1.3	0.63
AAA1 complex	4.0	0.25	1.0	1.01
AAA1 complex	2.5	0.25	0.9	1.63
AAA1 complex	2.0	0.25	0.8	3.18

Table 3 AAAs with applied surface mesh sizes

Models	Max element size in mm	Min. element size in mm	Mean element size in mm	Total number of FE element in million
AAA1 complex	2.5	0.25	0.9	1.63
AAA1 simplified	2.5	0.25	0.9	1.42
AAA2 complex	1.7	0.25	0.6	1.07
AAA2 simplified	1.7	0.25	0.6	0.72

3.2.2 Blood Model

Blood is assumed to be an incompressible and a homogeneous fluid that exhibits non-Newtonian shear thinning behavior for very small strain rates. Whereas for large strain rates, the blood behaves as a Newtonian fluid where viscosity remains constant [8, 14]. Validation has been carried out to observe differences when considering blood either as a Newtonian or a Non-Newtonian fluid model. The Newtonian blood model with a constant viscosity was implemented [4]. A simplified Non-Newtonian blood model based on Carreau power law, is presented in Eqs. (8) and (9),

$$\eta = \eta_\infty + (\eta_0 - \eta_\infty) f(\dot{\gamma}) \tag{8}$$

$$f(\dot{\gamma}) = \frac{1}{1 + \lambda \dot{\gamma}} \tag{9}$$

where η_0 is low shear rate asymptotic viscosity value 5.18 MPa, η_∞ is the high shear rate asymptotic viscosity, $\dot{\gamma}$ is the shear strain rate and λ is the time constant at 4.84 s [4]. The results presented in Fig. 10, highlight the difficulty in distinguishing the differences in the wall shear measured at the two points near neck region and bifurcating region of aneurysm for one cycle. Absence of differences is due to large strain rate correlated to high velocity in the arterial structures. And hence in the further simulations we have adopted constant viscosity for the blood model. The density of the blood was taken as 1,060 kg/m^3 [4].

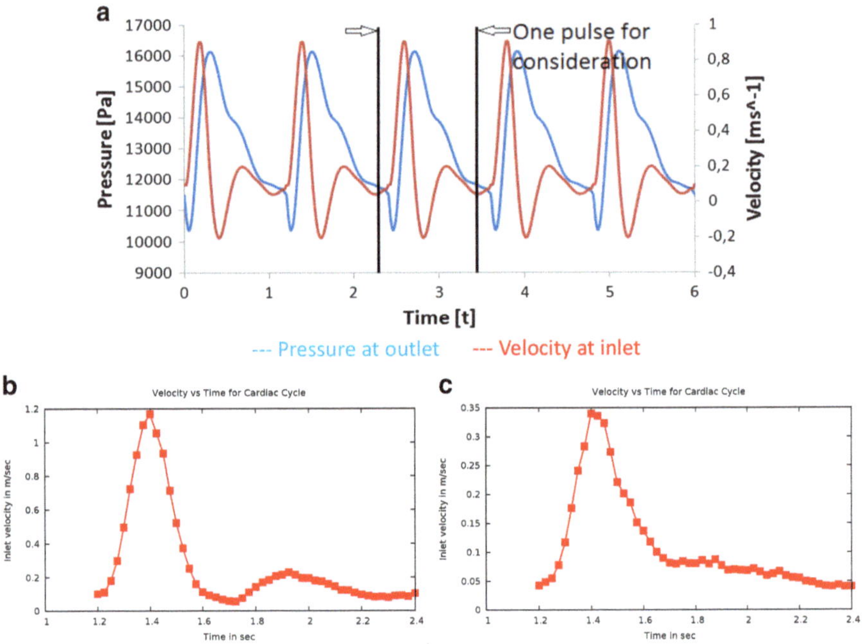

Fig. 9 Pressure and velocity wave forms applied at outlet and inlet of AAA models for one cardiac cycle of 1.2 s: (**a**) inlet velocity and outlet pressure profiles, (**b**) Inlet velocity at AAA2 simplified, (**c**) inlet velocity at AAA1 simplified

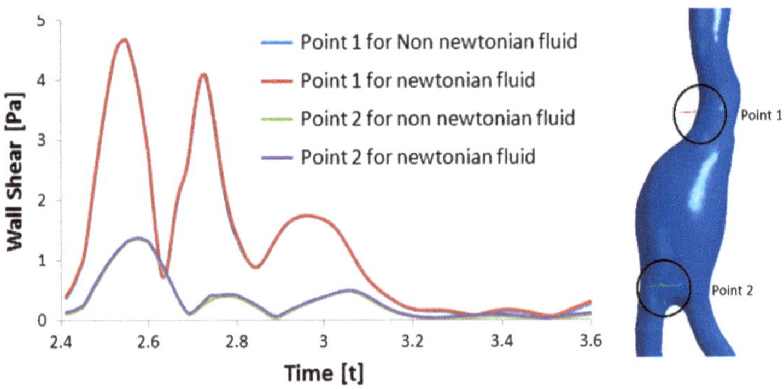

Fig. 10 Wall shear at two distal points for Newtonian and Non-Newtonian fluid

4 Results

Simulations were carried out by the `Starccm+`[3] solver for five cardiac cycles and result for each single cycle (1.2–2.4 s) are visualized with the help of Paraview. Two PSGs and their morphologically different geometries were analyzed during critical phases of a cardiac cycle like peak systole and diastole to demonstrate the pathological flow differences and wall shear differences. Velocity distributions, whirls and re-circulation zones, are presented on the cross sectional planes (plane1 and plane2) for both complex and simplified model of AAA to indicate the gradients of flow field. Results are presented at two critical phases of the cardiac cycle; peak systole at 1.4 s and diastole point at 1.6 s. The planes at which the flow variables are illustrated can be seen in Fig. 11. The pressure distribution in domains throughout the cardiac cycle remains similar. However, flow separation and re-circulation zones in both the cases vary throughout the cycle which is clearly visible in Figs. 12 and 13. During the systolic phase when the velocity in the domain is at peak it is interesting to note in Fig. 13, the vortex is still being generated in aneurysm sac at different locations. Reverse flow can be observed during the diastolic phase where the intensity of the flow reduces which results into a formation of whirls and stagnation points at different locations. The magnitude of the re-circulation zones is different and velocity gradient near wall tends to differ in both sets of AAAs.

The wall shear along with normal pressure force in aneurysm sac could be a critical factor in determining the localized high stress region in an arterial wall. In both AAA cases, the CM experiences wall shear on larger circumferential area of aneurysm sac compared to SM at peak systolic phase as shown in Fig. 14. These

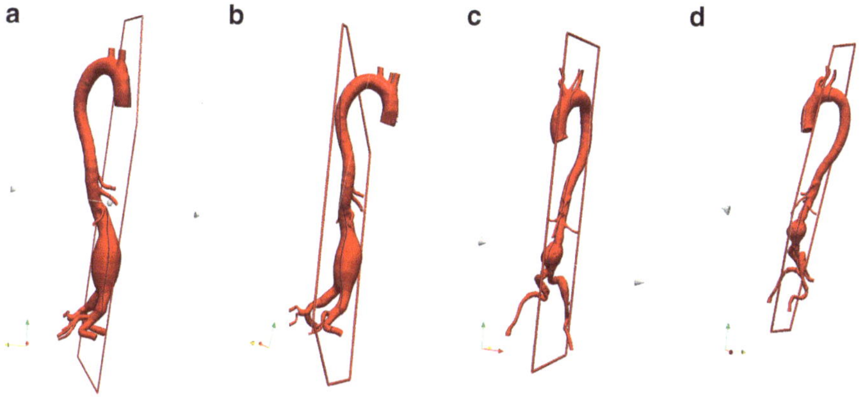

Fig. 11 Position of two planes on AAA models: **(a)** plane1 AAA1 **(b)** plane2 AAA1 **(c)** plane1 for AAA2 **(d)** plane2 for AAA2

---

[3]http://www.cd-adapco.com.

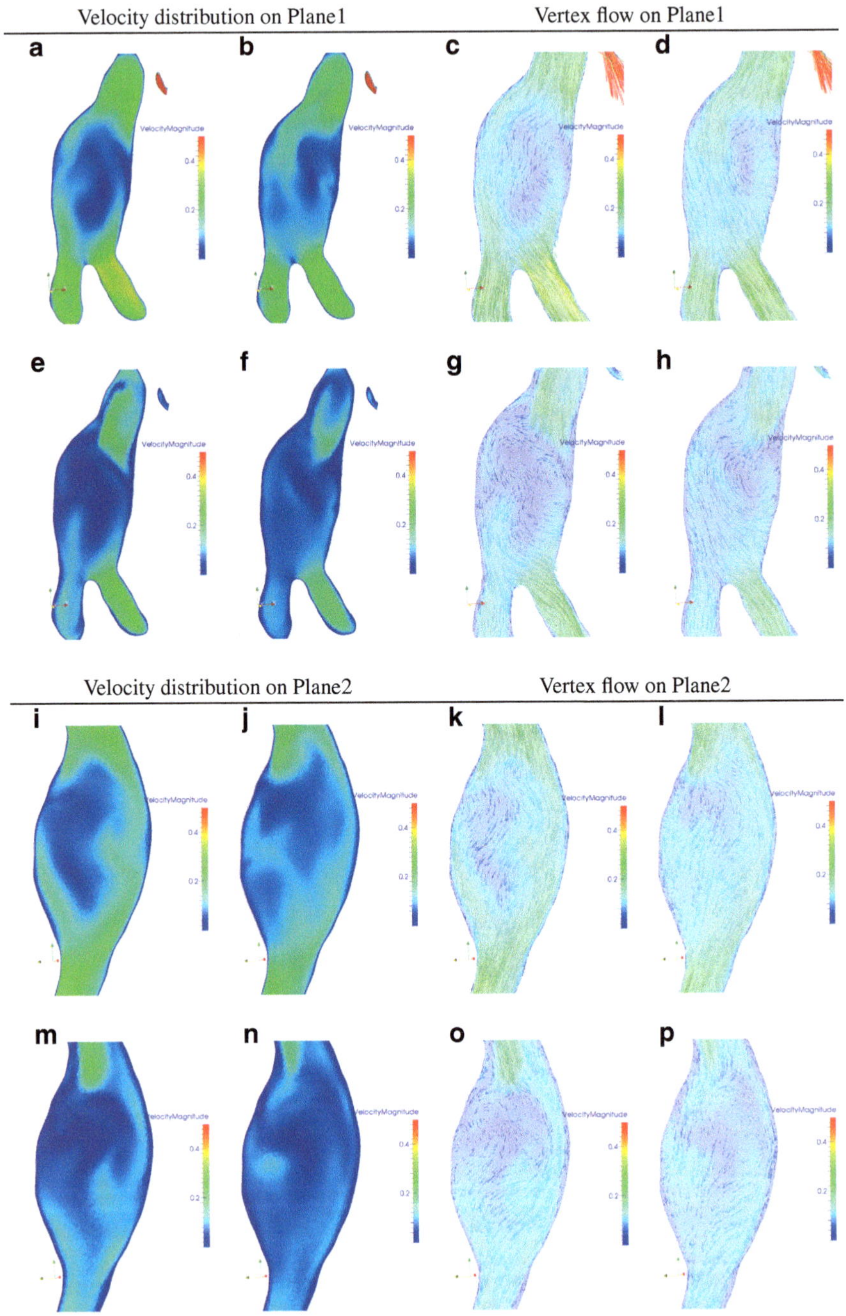

Fig. 12 Velocity distribution and re-circulation zones on two planes in SM and CM of AAA1: (**a**) CM at 1.4 s, (**b**) SM at 1.4 s, (**c**) CM at 1.4 s (**d**) SM at 1.4 s, (**e**) CM at 1.65 s, (**f**) SM at 1.65 s, (**g**) CM at 1.65 s, (**h**) SM at 1.65 s, (**i**) CM at 1.4 s, (**j**) SM at 1.4 s, (**k**) CM at 1.4 s, (**l**) SM at 1.4 s, (**m**) CM at 1.4 s, (**n**) SM at 1.4 s, (**o**) CM at 1.4 s, (**p**) SM at 1.4 s

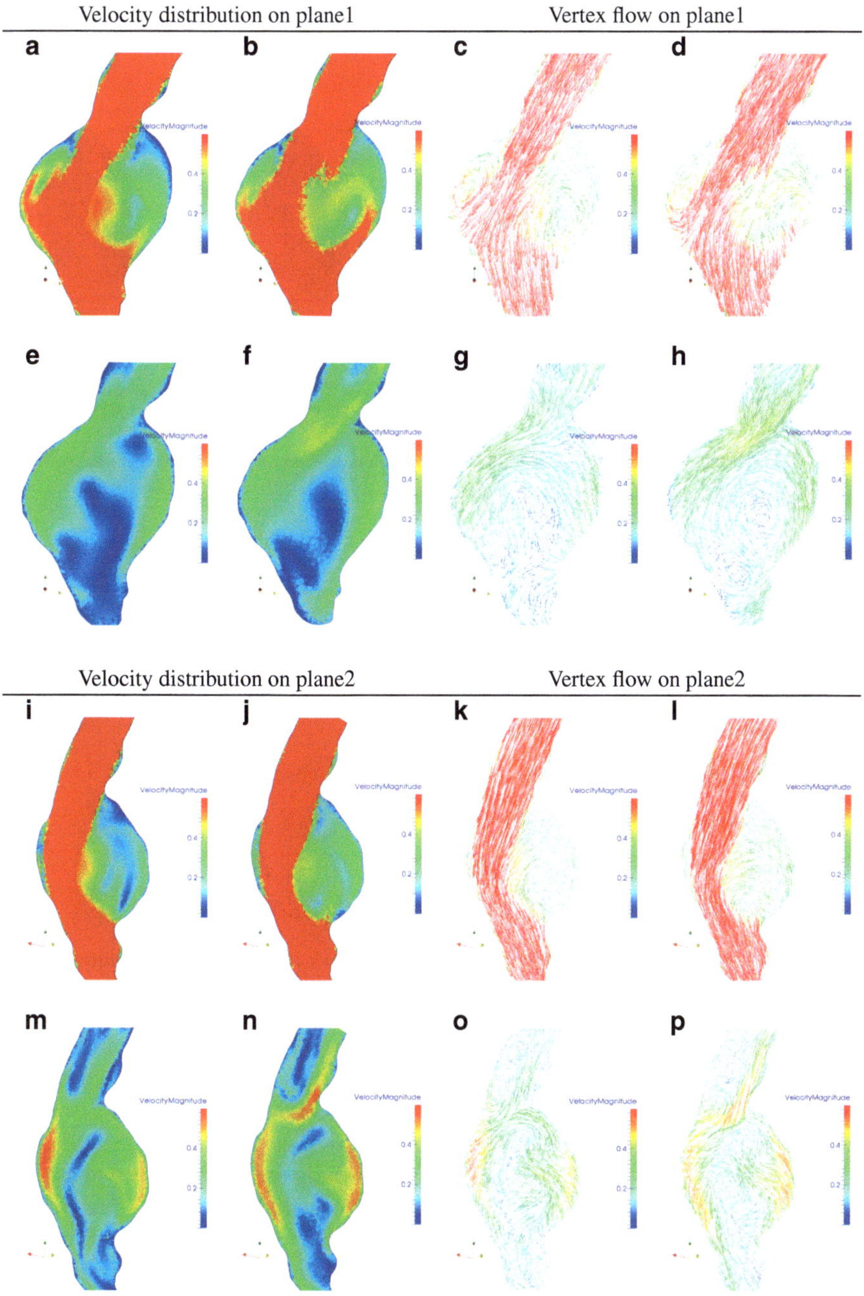

Fig. 13 Velocity distribution and re-circulation zone on Plane2 in CM and SM of AAA2: (**a**) CM at 1.4 s, (**b**) SM at 1.4 s, (**c**) CM at 1.4 s, (**d**) SM at 1.4 s, (**e**) CM at 1.65 s, (**f**) SM at 1.65 s, (**g**) CM at 1.65 s, (**h**) SM at 1.65 s, (**i**) CM at 1.4 s, (**j**) SM at 1.4 s, (**k**) CM at 1.4 s, (**l**) SM at 1.4 s, (**m**) CM at 1.4 s, (**n**) SM at 1.4 s, (**o**) CM at 1.4 s, (**p**) SM at 1.4 s

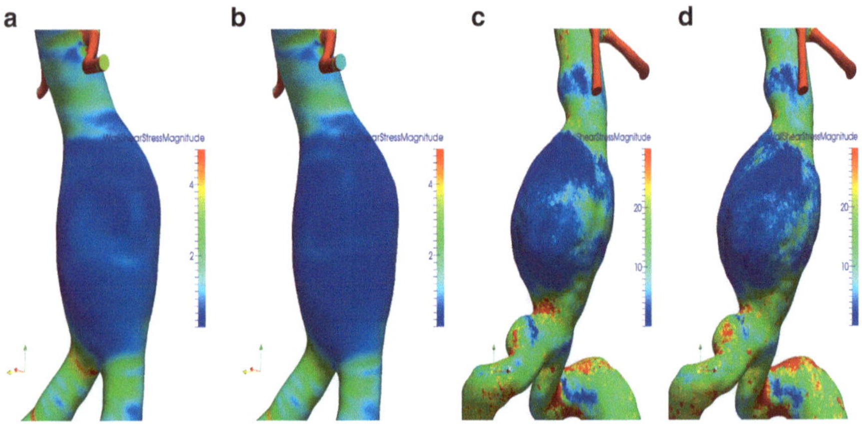

Fig. 14 Wall shear distribution in CM and SM of AAAs: (**a**) CM of AAA1, (**b**) SM of AAA1, (**c**) CM of AAA2, (**d**) SM of AAA2

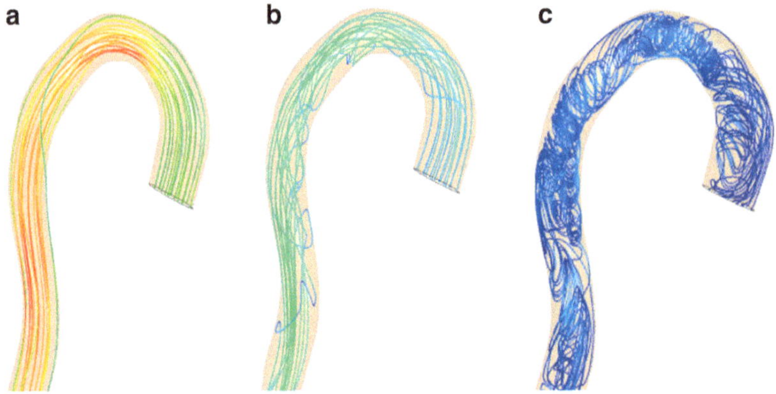

Fig. 15 Helical flow developing in aortic arc: (**a**) flow at 1.4 s, (**b**) flow at 1.6 s, (**c**) flow at 1.8 s

differences are less relevant in AAA1 case as aneurysm sac is much larger with maximum diameter exceeding 5 cm which triggers the local deceleration of flow.

Figure 15 gives more ideas about the rotational flow being initiated from the aortic arc. Systolic peak where the velocity is highest we observe that the flow behaves linearly and as the cycle progresses towards the late diastolic phase, see Fig. 15b, c, the radial rotations of the fluid particles with high degree of turbulence starts showing up. This natural phenomena arises due to morphological differences and cannot be achieved by any theoretical assumptions applied on any SM with restricted computational domains.

Even though the global pressure distribution along with pressure drop in the CM and SM of AAAs are quite similar to each other we have observed pathological differences. These strong secondary flow differences arise as a result of varying complexity of the geometry. The arc of the aorta plays a significant role in governing the fluid flow downstream into the aneurysm. The aortic arc naturally induces radial rotation in the flow like a spiral flow by it's natural morphology. This spiral flow may be caused by ejection of the flow from heart at a particular angle and the basic topology of vascular structure. The topology of both the AAAs presented here seems to be helical in nature downstream like a radial twist from the beginning of the arc. Along with the arc, neck angle at which flow is entering the aneurysm sac also plays important role in governing flow but this angle remains the same for both SM and CM. These results shows us that geometry of vascular structures of each patient varies significantly. And how flow behaves differently on including aortic structures which in general are neglected for computational studies in AAAs. This study indicate that the geometric development of vascular structures has a major role on the resulting flow patterns which may affect the structural dynamics of the arterial wall. Therefore, it is necessary to model local haemodynamic phenomena while considering realistic geometric models of vascular structures.

There are certain limitations in this study and a few theoretical assumptions were taken into account. For example, patient specific inlet and outlet boundary conditions were not yet considered, thrombosis in aneurysm sac was excluded in the aneurysm sac and only fluid simulation results were presented neglecting the fluid structure interaction. Nevertheless the pathological differences due to morphological differences were presented. The next step of this study is to implement adaptive filters and include Fluid Structure Interaction considering the local varying thickness of the aortic structure.

References

1. Abdominal Aortic Aneurysm: http://www.medicinenet.com/abdominal_aortic_aneurysm/index.htm (2014)
2. Fillinger, M.F., Racusin, J., Baker, R.K., Cronenwett, J.L.: Anatomic characteristics of ruptured abdominal aortic aneurysm on conventional CT scans: implications for rupture risk. J. Vasc. Surg. **39**(6), 1243–1252 (2004)
3. Fillinger, M.F., Raghvan, M.L., Marra, S.P., Cronenwett, J.L., Kennedy, F.E.: In vivo analysis of mechanical wall stress and abdominal aortic aneurysm rupture risk. J. Vasc. Surg. **36**(3), 589–597 (2002)
4. Galdi, G.P., Rannacher, R., Robertson, A.M., Turek, S.: Hemodynamical Flows; Modelling, Analysis and Simulation. Birkhauser, Basel (2008)
5. Kniemeyer, H.W., Kessler, T., Reber, P.U., Ris, H.B., Hakki, H., Widmer, M.K.: Treatment of ruptured abdominal aortic aneurysm, a permanent challenge or a waste of resources? Prediction of outcome using a multi-organ-dysfunction score. Eur. J. Vasc. Endovasc. Surg. **19**(2), 190–196 (2000)
6. Lederle, F.A., Wilson, S.E., Johnson, G.R., Reinke, D.B., Littooy, F.N., Acher, C.W., Ballard, D.J., Messina, L.M., Gordon, I.L., Chute, E.P., Krupski, W.C., Busuttil, S.J., Barone, G.W., Sparks, S., Graham, L.M., Rapp, J.H., Makaroun, M.S., Moneta, G.L., Cambria, R.A., Makhoul, R.G., Eton, D., Ansel, H.J., Freischlag, J.A,. Bandyk, D.: Immediate repair compared with surveillance of small abdominal aortic aneurysms. N. Engl. J. Med. **346**(19), 1437–1444 (2002). www.nejm.org

7. Li, Z., Kleinstreuer, C.: A comparison between different asymmetric abdominal aortic aneurysm morphologies employing computational fluid–structure interaction analysis. Eur. J. Mech. B/Fluids **26**, 615–631 (2007)
8. Louiza, A.C.H.A.B., Mahfoud, M., Benhadid, S.: Numerical study of the non-newtonian blood flow in a stenosed artery using two rheological models: http://www.doiserbia.nb.rs/img/doi/0354-9836/2013%20OnLine-First/0354-98361300161A.pdf (2013)
9. Olufsen, M.S., Peskin, C.S., Pedersen, E.M., Kim, W.H., Nadim, A., Larsen, J.: Numerical simulation and experimental validation of blood flow in arteries with structured-tree outflow conditions. Ann. Biomed. Eng. **28**, 1281–1299 (2000)
10. Papaharilaou, Y., Ekaterinaris, J.A., Manousaki, E., Katsamouris, A.N.: A decoupled fluid structure approach for estimating wall stress in abdominal aortic aneurysms. J. Biomech. **40**, 367–377 (2007)
11. Perktold, K., Peter, R.O., Resch, M.M., Langs, G.: Pulsatile non-Newtonian blood flow in three-dimensional carotid bifurcation models: a numerical study of flow phenomena under different bifurcation angles. J. Biomed. Eng. **13**, 507–515 (1991)
12. Perktold, K., Peter, R.O., Resch, M.M.: Three-dimensional numerical analysis of pulsatile flow and wall shear stress in the carotid artery bifurcation. J. Biomed. Eng. **24**, 409–420 (1991)
13. Perktold, K., Hilbert, D.: Numerical simulation of pulsatile flow in a carotid bifurcation model. J. Biomed. Eng. **8**, 193–199 (1991)
14. Quemada, D.: Rheology of concentrated disperse systems III. General features of the proposed non-newtonian model. Comparison with experimental data. Rheol. Acta **17**(6), 643–653 (1978)
15. Santosa, F., Moysidis, T., Nowak, T., Heilmaier, C., Berg, C., Luther, B., Kröger, K.: Endovascular abdominal aneurysm repair: trends in Germany. Vasa **41**, 268-274 (2012)
16. Surgeon explains who needs screening, treatment: http://www.sciencedaily.com/releases/2014/07/140716112248.htm (2014)
17. Wolters, B.J.B.M., Rutten, M.C.M., Schurink, G.W.H., Kose, U., Hart, J.de, an de Vosse, F.N.: A patient-specific computational model of fluid–structure interaction in abdominal aortic aneurysms. J. Biomed. Eng. **24** 409–420 (1991)

MIX
Papier aus verantwortungsvollen Quellen
Paper from responsible sources
FSC® C105338

If you have any concerns about our products, you can contact us on
ProductSafety@springernature.com

In case Publisher is established outside the EU, the EU authorized representative is:
Springer Nature Customer Service Center GmbH
Europaplatz 3, 69115 Heidelberg, Germany

Printed by Libri Plureos GmbH
in Hamburg, Germany